AF855798

K9 Anatomy
Made Easy

Mind Maps, Mnemonics, Analogies, Audio QRs, Clinical, & Beyond

Prof. Dr. Fawzy A. Elnady
Professor of Anatomy and Embryology
Virginia Tech University
Virginia Maryland College of Veterinary Medicine
Department of Biomedical Sciences and Pathobiology

Part 1
Basic Concepts
& Forelimb

lulu
ZINSIDE BEAUTY
Safety & Longevity

Zinsidebeauty.com

Lulu.com

ISBN: 979-8-9883947-5-4

Dedication

To my wonderful family — my beloved wife, Dr. Asmaa Elkordy, whose love and support inspire me every day; and our children, Yusuf, Sondos, and Maryam, who bring joy and meaning to my life.

Preface

Anatomy is the foundation of veterinary medicine, yet for many students, it often appears as a complex, daunting subject. As an educator and veterinary anatomist for decades, I've seen firsthand the struggles that come with mastering the intricate structures and functions that define the animal body. The challenge lies not only in memorizing a vast amount of information but also in understanding the relationships and connections between various anatomical systems.

"Veterinary Anatomy Made Easy: Mind Maps, Mnemonics, Analogies, Audio QRS, Clinical & Beyond" was born out of the desire to simplify this process. This book aims to break down the complexities of veterinary anatomy into digestible, accessible concepts using powerful learning tools: mind maps, mnemonics, and analogies. By connecting anatomical structures to everyday experiences and visual aids, I hope to make anatomy not just easier to learn, but also more engaging and memorable.

Throughout this book, Part I - Thoracic limb, you'll find carefully crafted mind maps that illustrate key anatomical concepts and their interrelationships, making it easier to visualize and understand the material. The mnemonics are designed to help you remember crucial details with minimal effort, and the analogies provide a bridge between the unfamiliar language of anatomy and everyday experiences.

My goal with this book is not only to help students navigate the complexities of veterinary anatomy but also to inspire a deeper appreciation for the beauty and intricacy of the animal body. Whether you're a veterinary student, a professional, or simply an animal enthusiast, this book is designed to be a useful, practical resource for learning and applying veterinary anatomy in real-world settings.

I hope that by the end of this book, you'll find yourself more confident in your understanding of the canine thoracic limb anatomy, equipped with the tools to approach this challenging subject with ease and enthusiasm.

Dr. Fawzy Elnady
Professor of Anatomy and Embryology
Virginia-Maryland College of Veterinary Medicine
Virginia Tech University

How to use this book

To get the most out of this book, follow these simple guidelines:

1. Learn visually and audibly, start with the mind maps and scan the QR code to access the audio narration.

Each chapter has a mind map that visually outlines the key anatomical concepts and their interconnections. These mind maps serve as a roadmap, helping you see the "big picture" before diving into the details. Use the mind maps as a reference point throughout your studies. They are great for quickly reviewing a topic and seeing how various anatomical structures relate to one another.

2. Use the Mnemonics for Easy Recall

Throughout the book, you'll find mnemonics designed to make memorization easier. These memory aids are especially useful for recalling complex terms, lists, or sequences.

When encountering a mnemonic, take time to learn it. Repeating it aloud or writing it down will help solidify it in your memory.

Feel free to create your own mnemonics based on the examples provided. This personal touch can further enhance your understanding and retention.

3. Leverage Analogies to Deepen Understanding

To help you grasp difficult concepts, the book also includes analogies that link anatomical structures to everyday objects or experiences. These analogies simplify complex ideas by relating them to something familiar, making it easier to visualize and understand.

As you read, try to relate the analogies to your own experiences or find new ways to connect the concepts to things you already know.

4. Interactive Learning

To enhance your learning experience, this book encourages active engagement with the material. Try the following:

Draw your own mind maps based on what you've learned to reinforce the connections between concepts.

Quiz yourself using the mnemonics and analogies. Write down what you remember before checking the answers in the text.

Discuss the material with classmates or colleagues to solidify your understanding through conversation.

5. Go Beyond the Basics

While this book focuses on simplifying veterinary anatomy, feel free to explore additional resources or consult with professors and professionals to expand your knowledge.

By using this book as a practical, interactive tool, you'll not only learn veterinary anatomy more effectively but also enjoy the process. Remember, understanding the animal body is both a science and an art, and with the right tools, you can master this subject and apply it with confidence in your career.

Happy studying, and may this book be a valuable companion on your journey through veterinary anatomy!

Table of Contents

Abbreviations

Pr	*process*	*N*	*nerve*	*jt* *jts*	*joint* *joints*
DE	*Digital Extensor*	*Nn*	*nerves*	*tub*	*tuberosity*
DF	*Digital flexor*	*Ln* *LC*	*lymph node* *lymph Center*	*tuberc*	*tubercle*
dist	*distal*	*w/*	*with*	*for*	*foramen*
prox	*proximal*	*cranial*	*cr*	*art*	*articulate or articular*
br *brs*	*branch* *branches*	*caudal*	*cd*	*gro*	*groove*
A *Aa*	*artery* *arteries*	*Med*	*medial*	*dig*	*digital*
Lig	*ligament*	*Lat*	*lateral*	*Rt.*	*right*
V *Vv*	*vein* *Veins*	*mid*	*middle*	*Lt.* *M* *Mm*	*left* *Muscle* *Muscles*

CHAPTER 1

Anatomical and Directional Terms

ANATOMY • ORIENTATION • POSITION • PLANES • DIRECTIONS

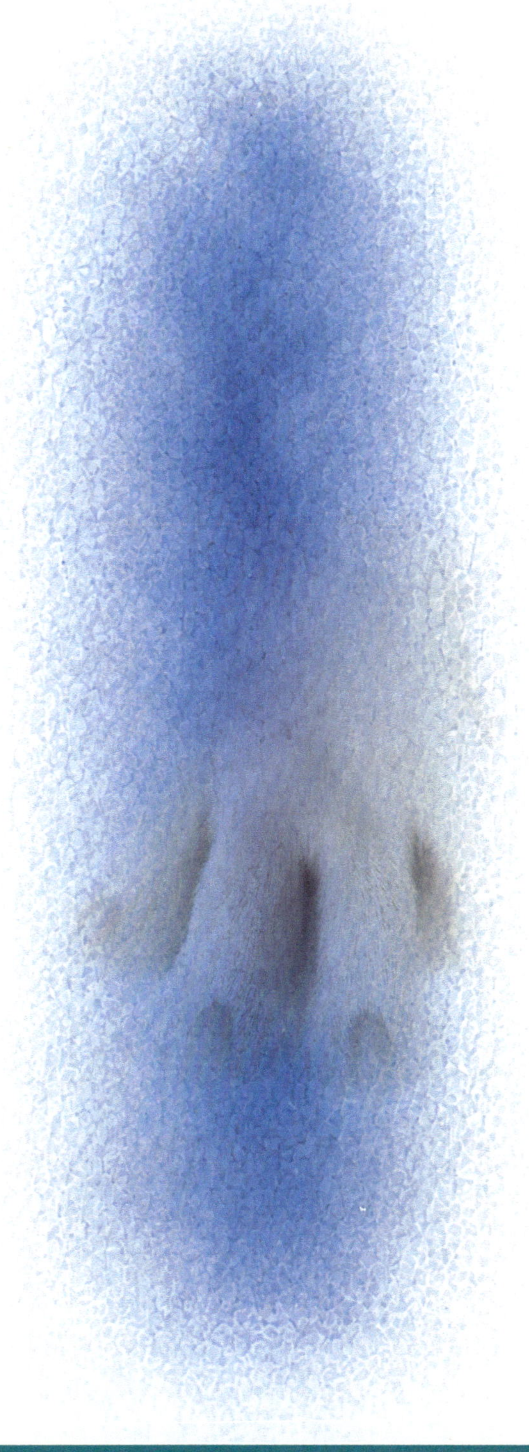

Anatomical terms

- **Planes**
  - median
  - sagitttal / paramedian
  - transverse
  - dorsal
- **Directional**
  - dorsal & palmar / plantar
  - proximal & distal
  - rostral & caudal
  - anterior & posterior
  - superior & inferior
  - head & eye
  - cranial & caudal
  - medial & lateral
  - dorsal & ventral
  - superficial & deep
  - axial & abaxial
- **Movement**
  - flexion & extension
  - hyperextension
  - adduction (+) & abduction
  - protraction & retraction
  - elevation & depression
  - pronation & supination
  - inversion & eversion
  - rotation
  - circumduction

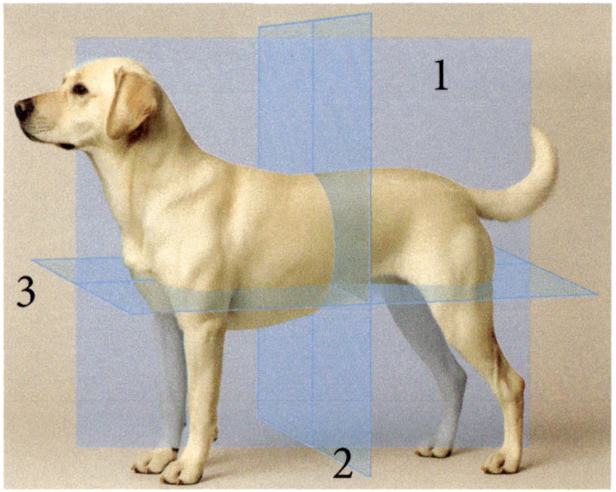

posterior
anterior
The water level
is a dorsal plane.

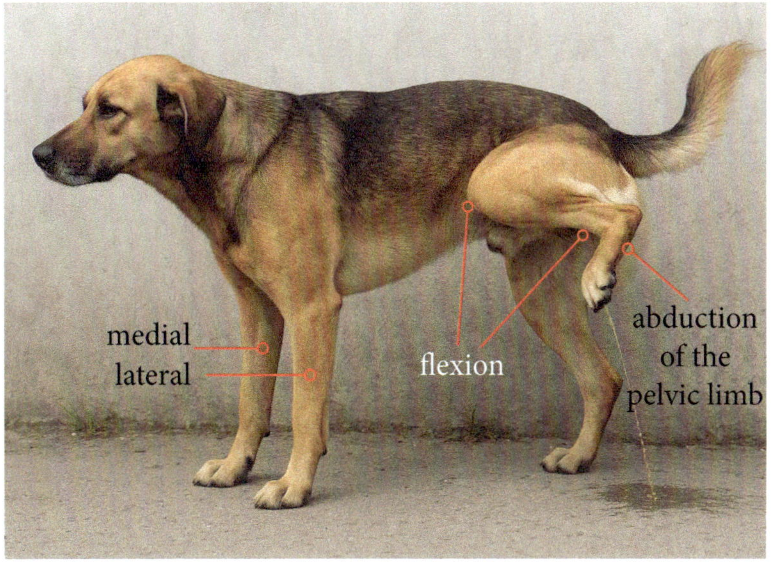
medial
lateral
flexion
abduction
of the
pelvic limb

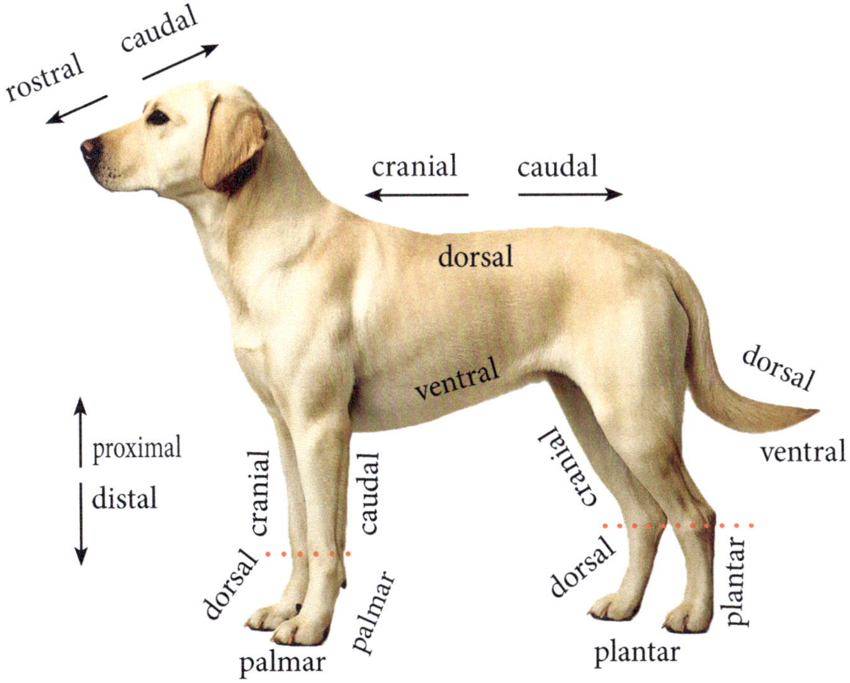
rostral
caudal
cranial
caudal
dorsal
ventral
dorsal
ventral
proximal
distal
cranial
caudal
cranial
dorsal
palmar
palmar
dorsal
plantar
plantar

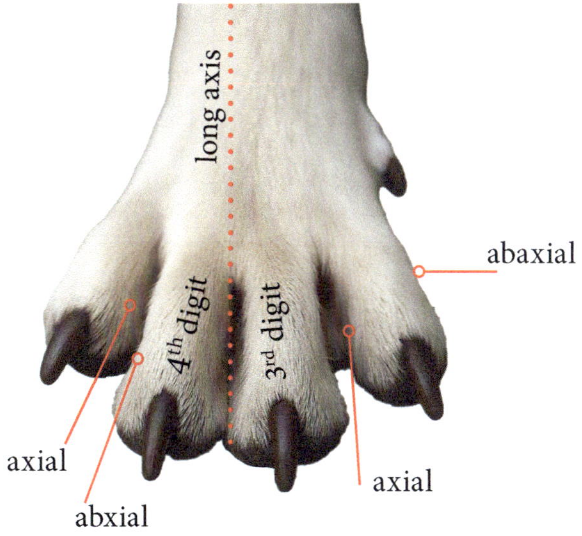
long axis
abaxial
4th digit
3rd digit
axial
abxial
axial

CHAPTER 2

The Skeletal Framework

SKELETON OVERVIEW • BONE TYPES • PROJECTIONS • DEPRESSIONS • OPENINGS

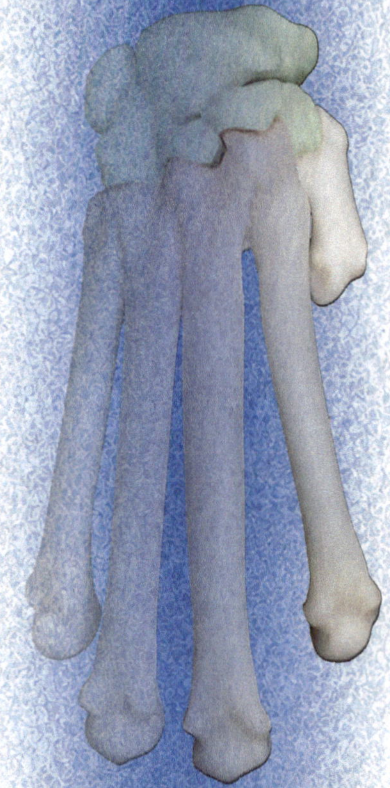

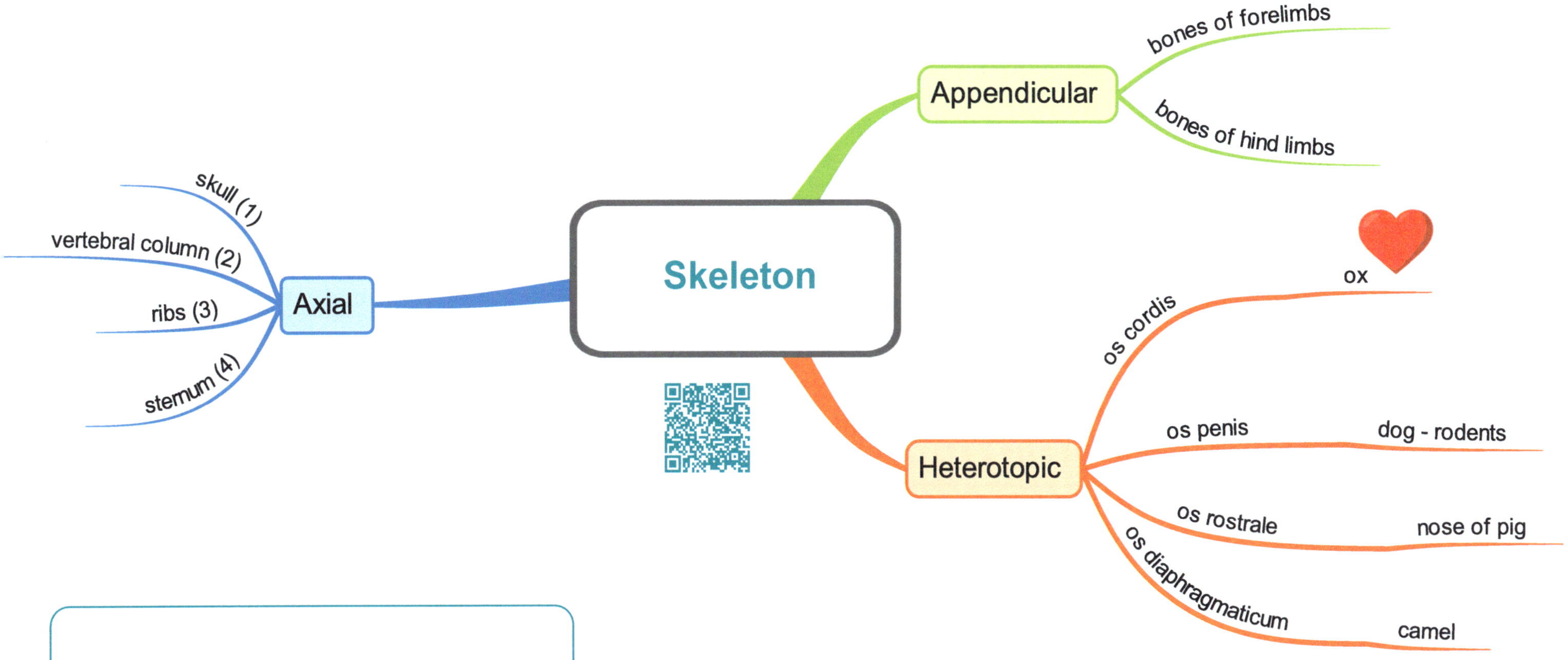

***Notice:** The axial skeleton includes all bones in the body except those in the limbs.*

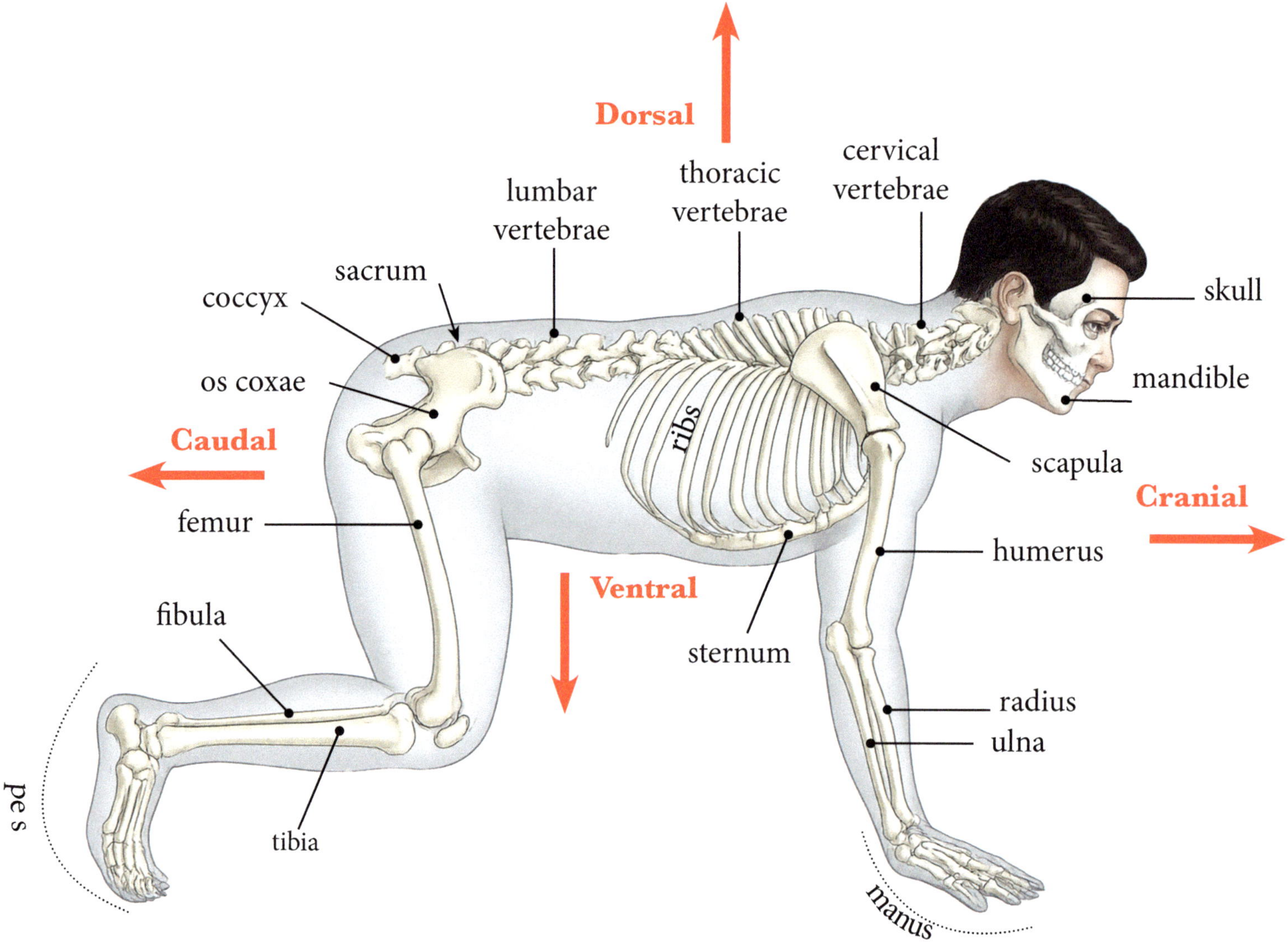

Human skeleton

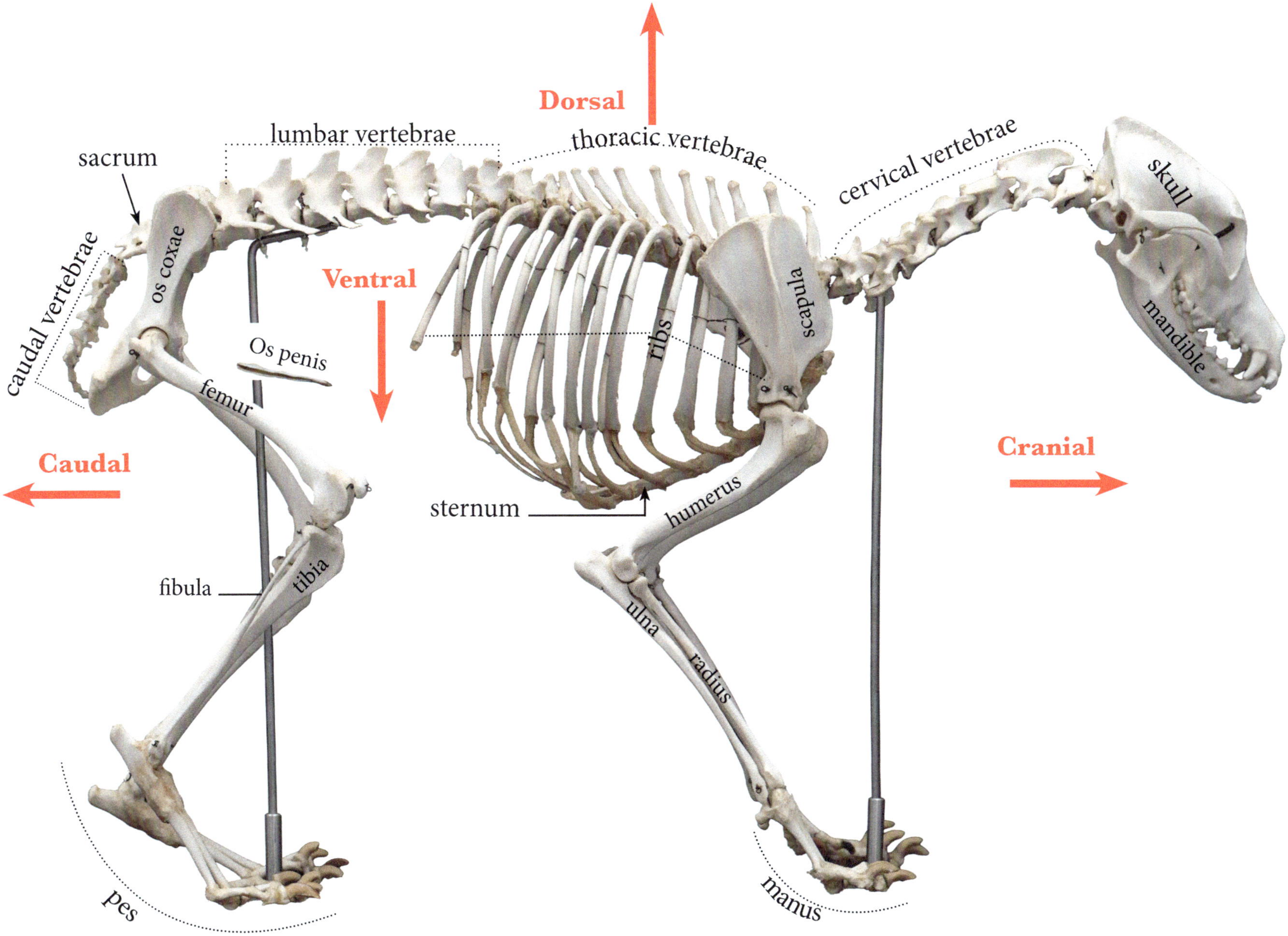

Skeleton of a dog

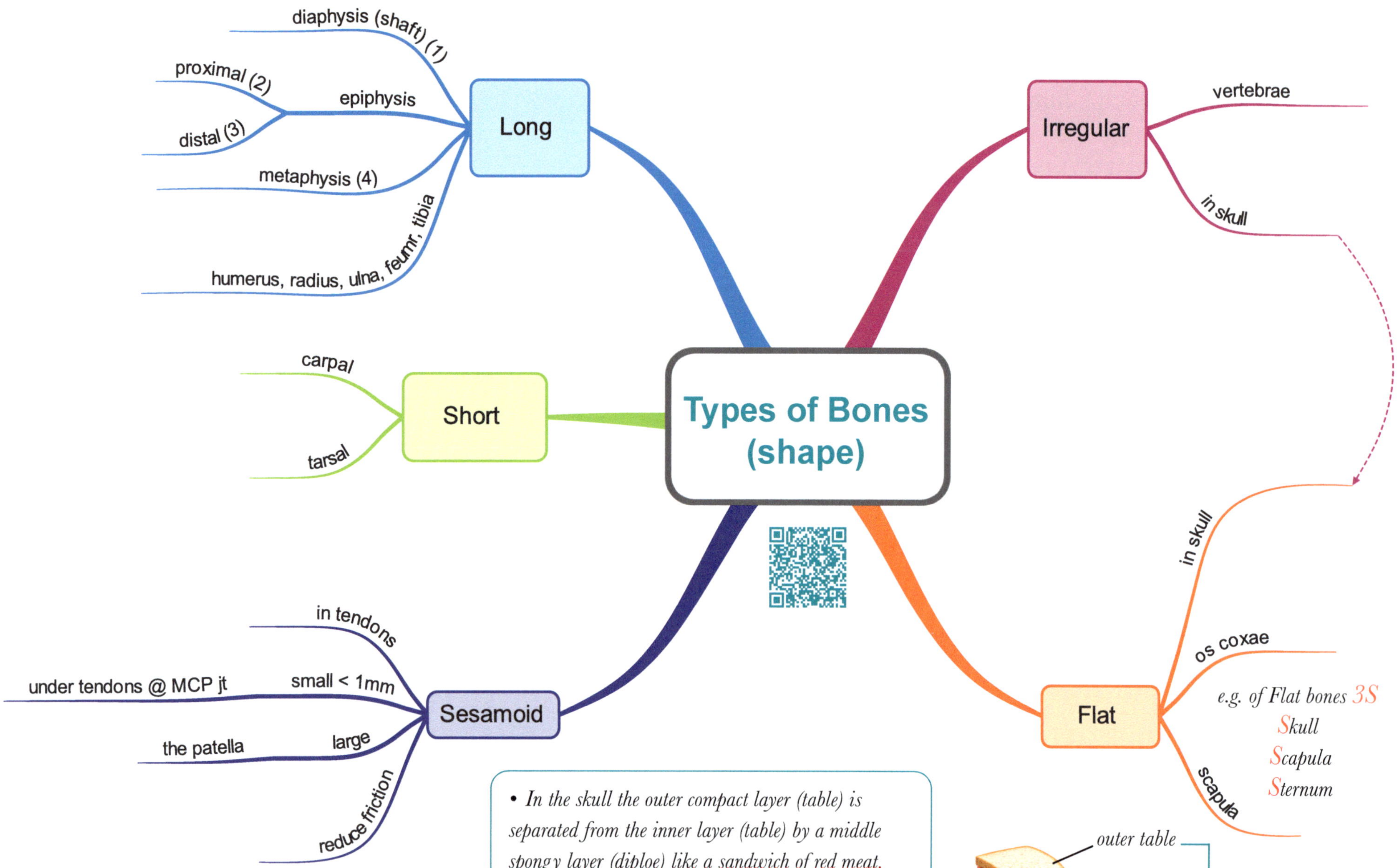

- *In the skull the outer compact layer (table) is separated from the inner layer (table) by a middle spongy layer (diploe) like a sandwich of red meat.*
- *In pneumatic bones the diploe disappear and replaced by air forming paranasal sinuses.*

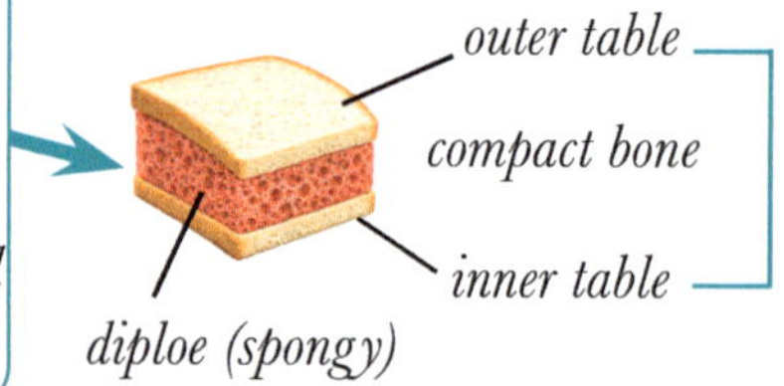

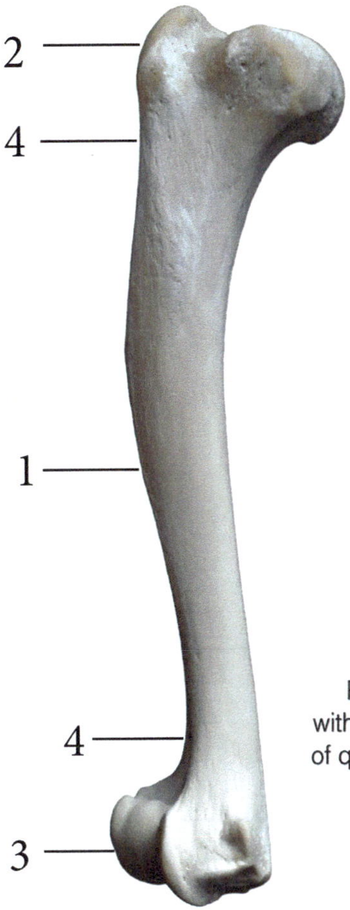

Humerus (long bone)

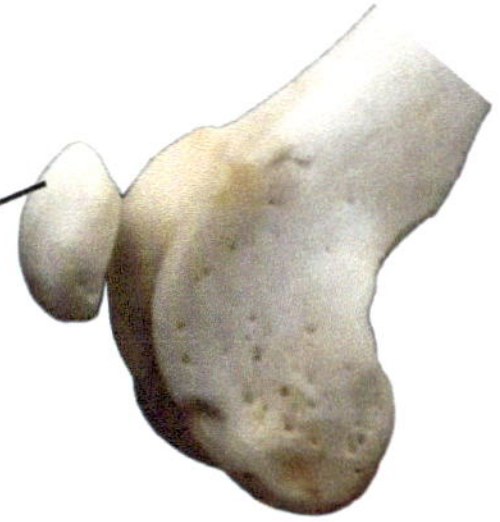

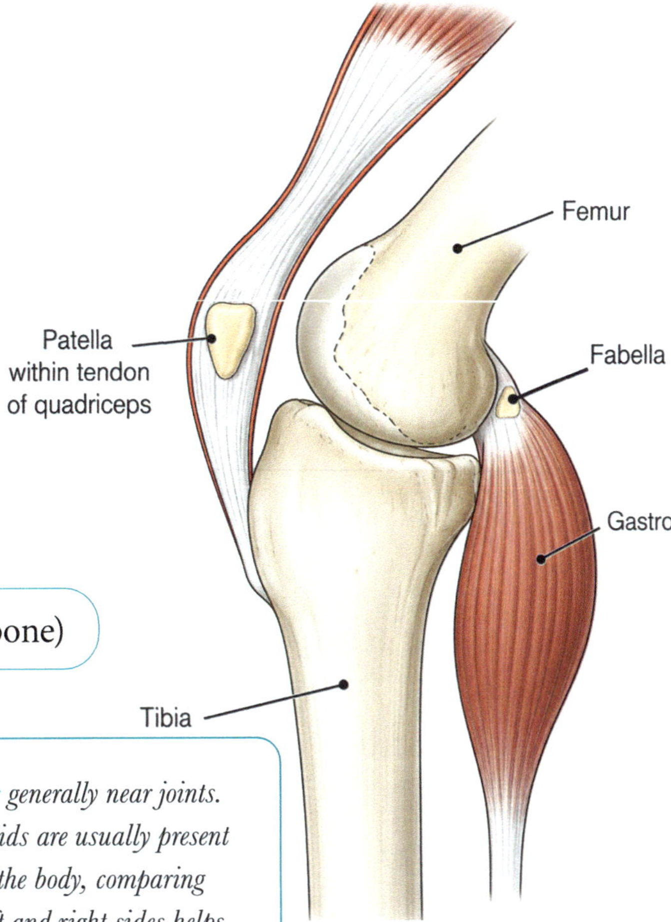

Sesamoids

Sesamoid bones generally near joints. Since sesamoids are usually present on both sides of the body, comparing X-rays of the left and right sides helps distinguish them from fractures.

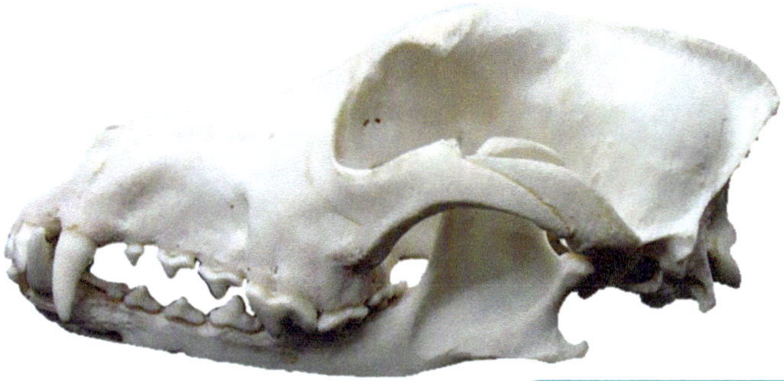

Skull (flat & irregular)

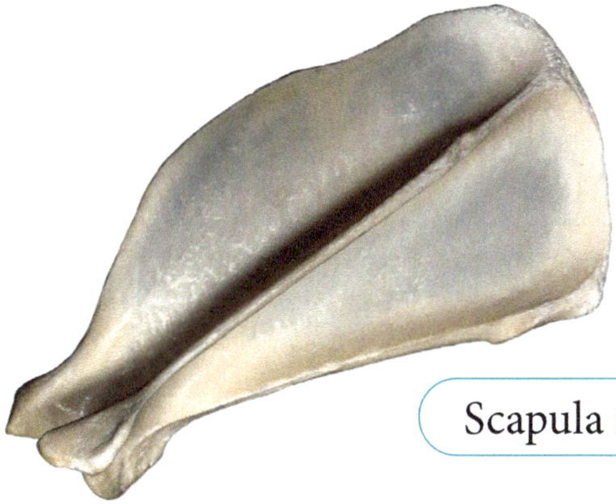

Scapula (flat bone)

Carpal bones
(short bones)

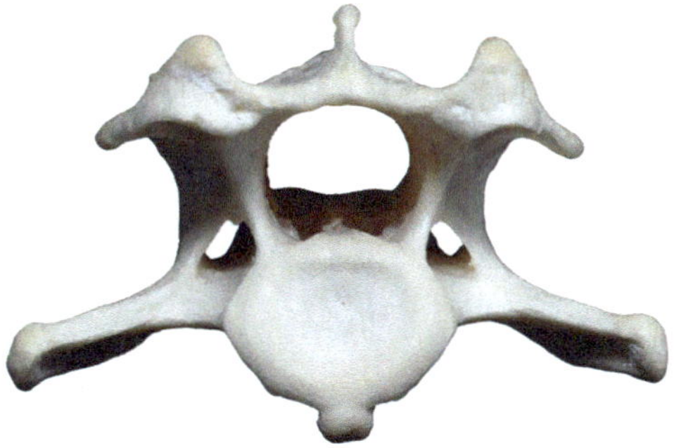

Cervical vertebra (irregular)

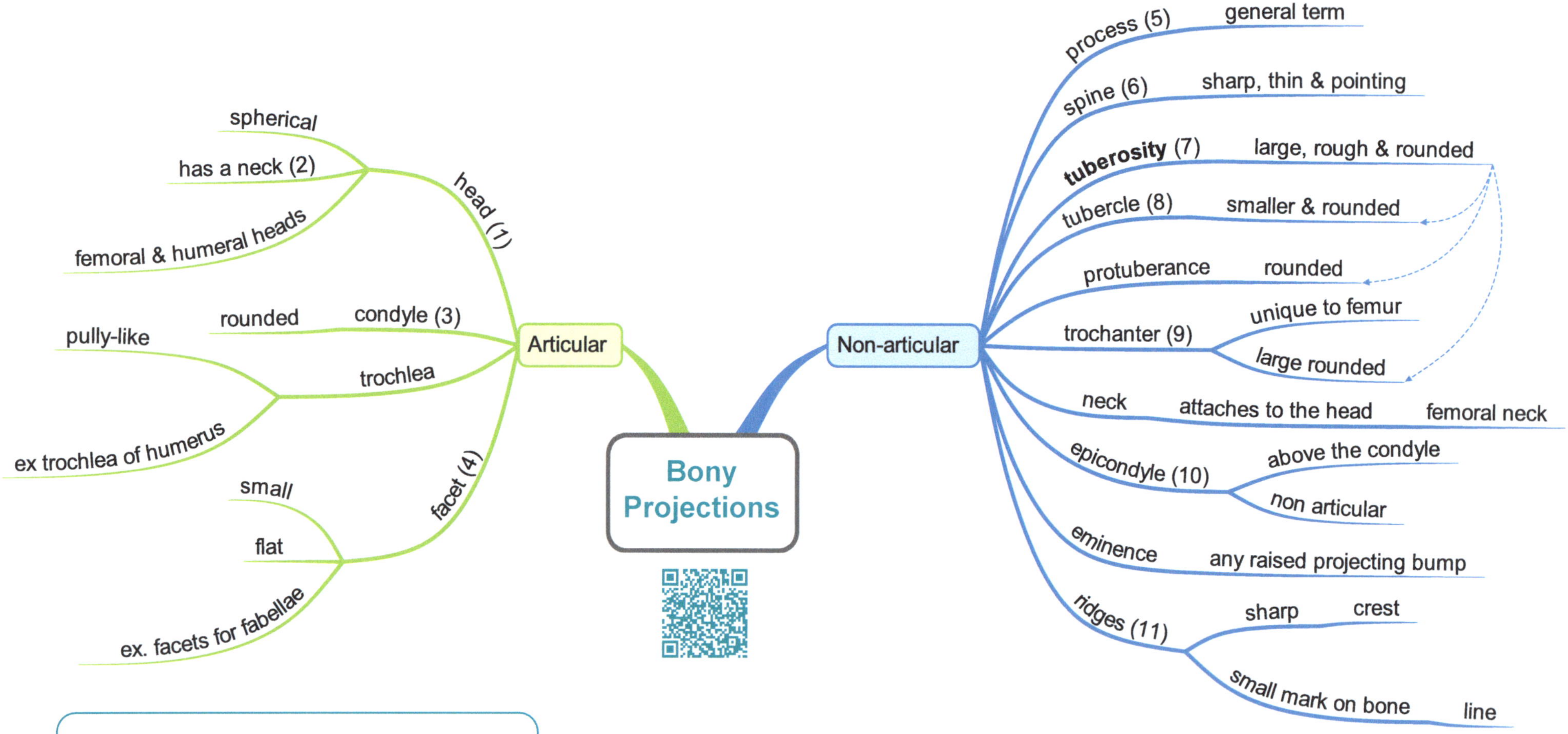

Although tuberosity, tubercle, and tuber are slightly different, they may be used interchangeably at this level.

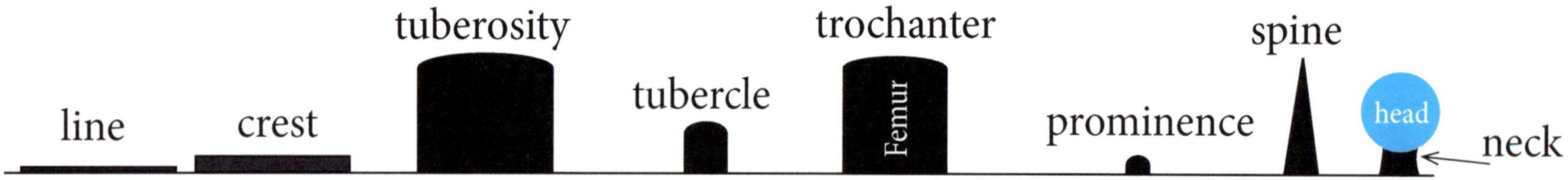

Non-articular bony projections

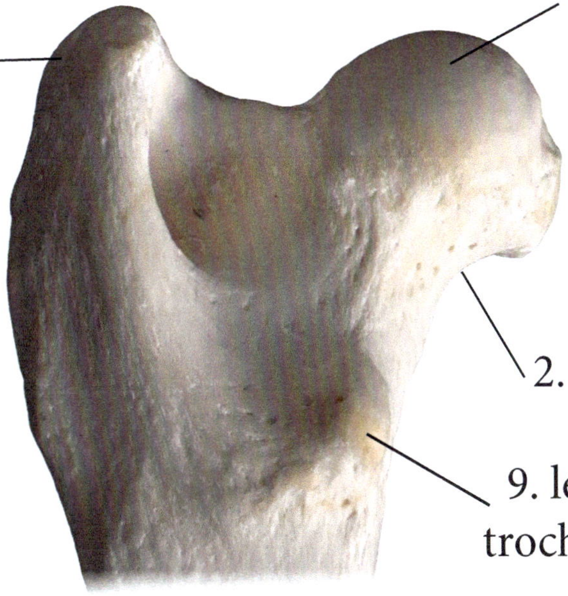

Proximal femur

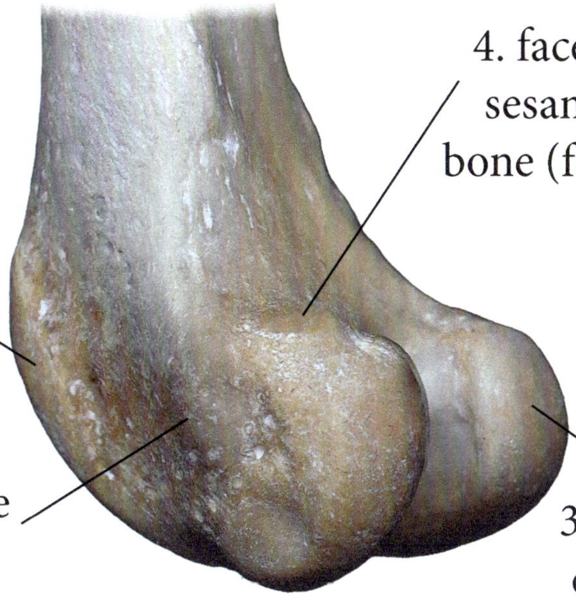

Distal femur

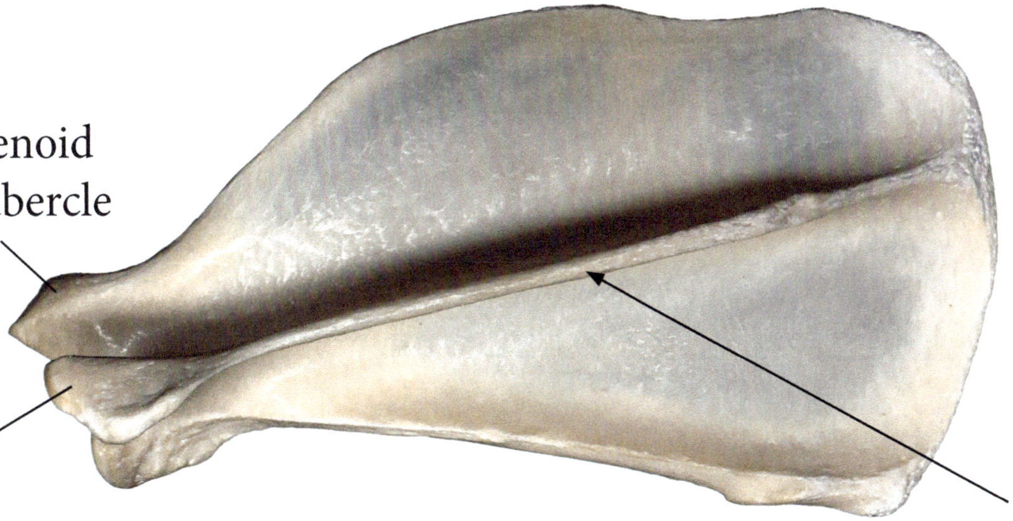

Scapula- lateral view

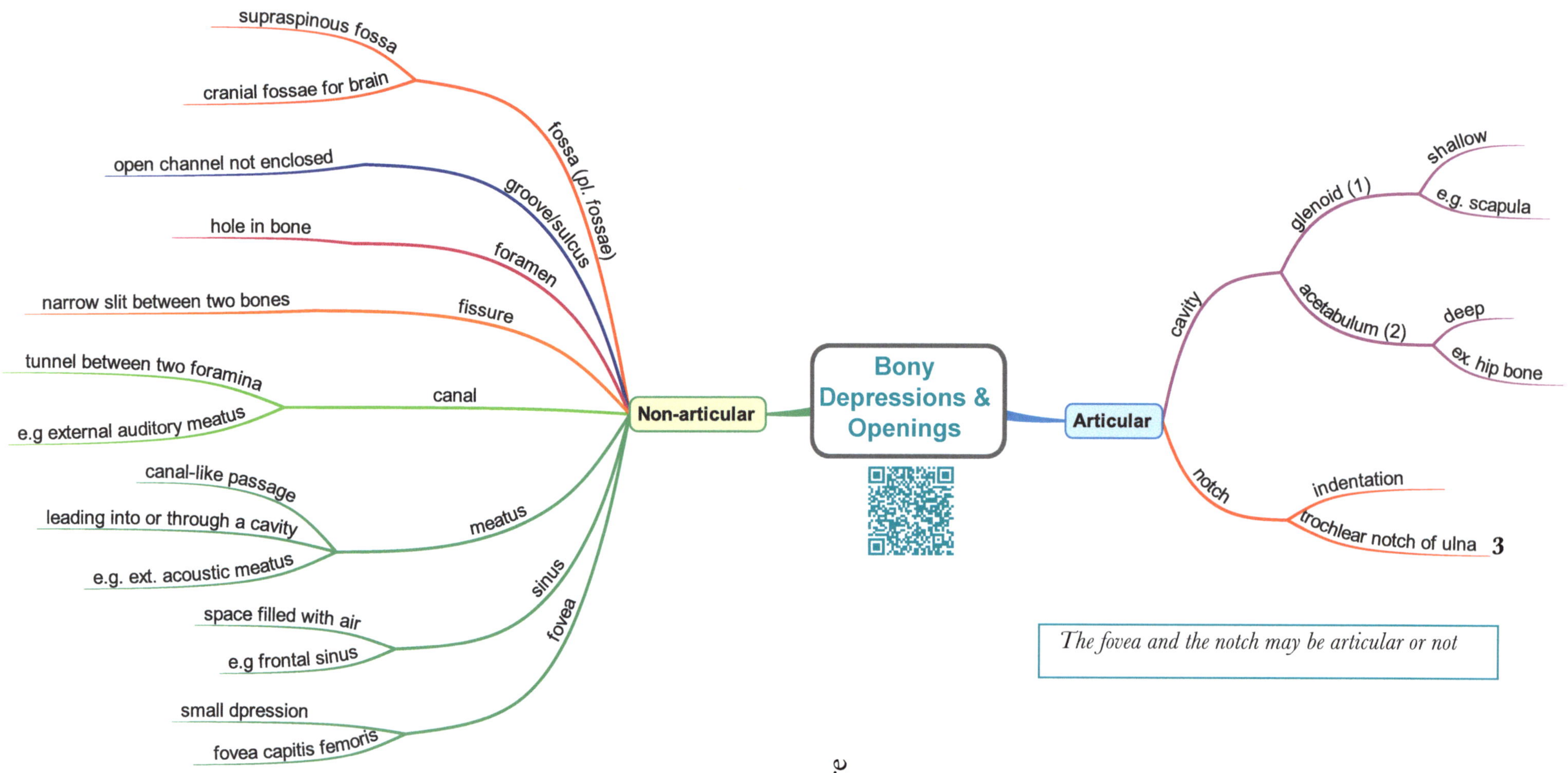

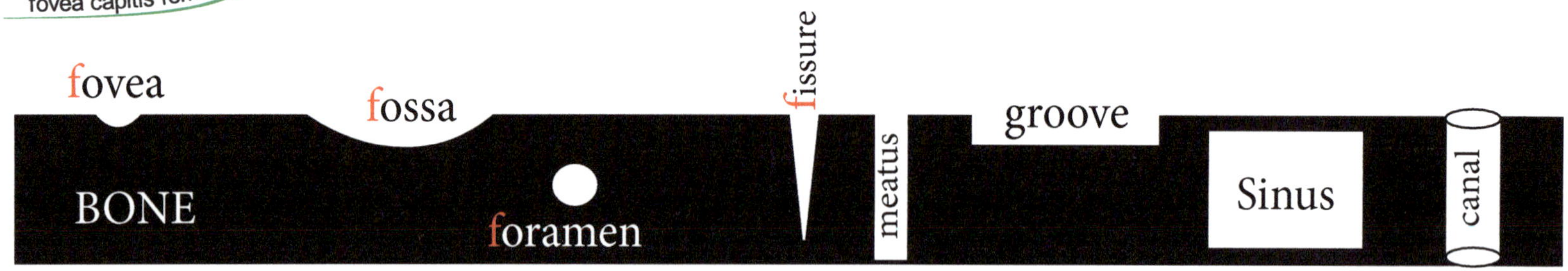

Non-articular bony depressions

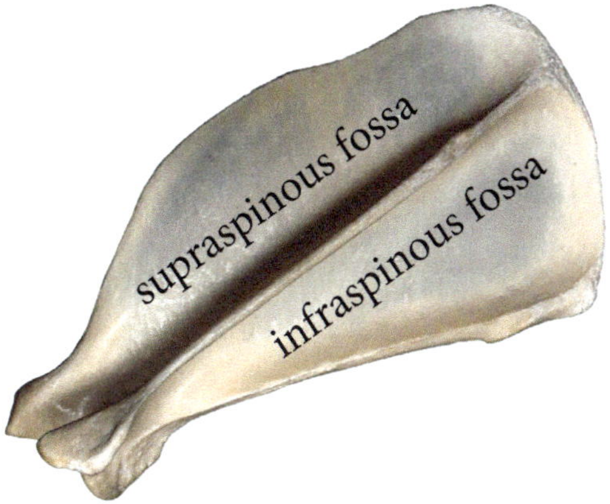

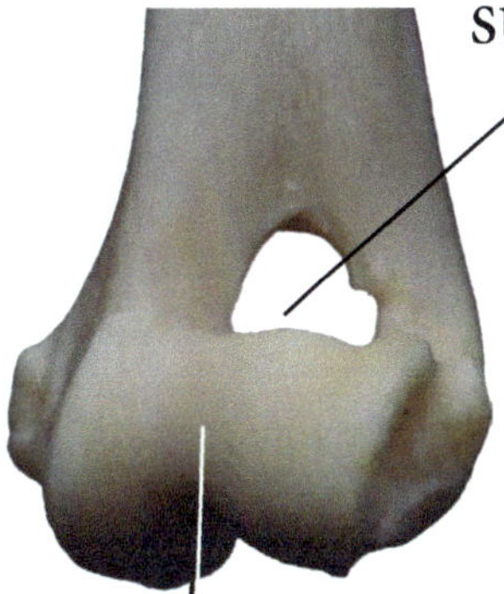

Bony projection

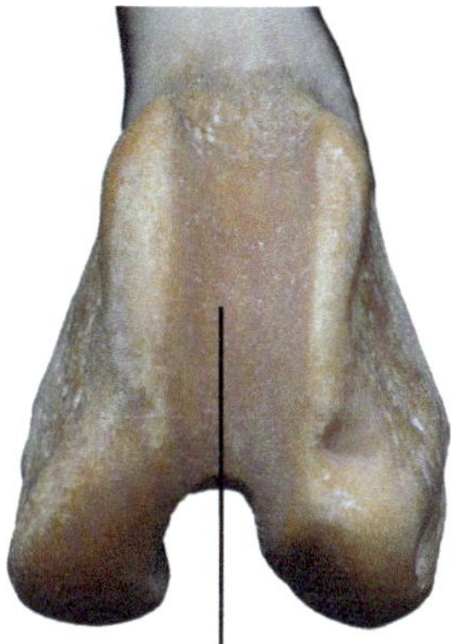

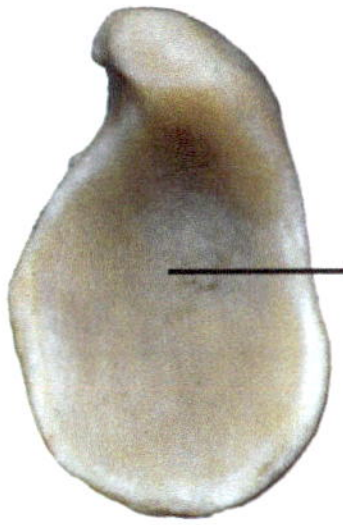

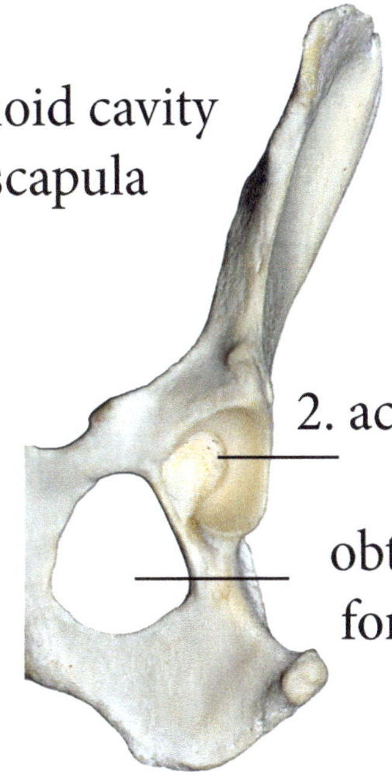

hip bone (os coxae)

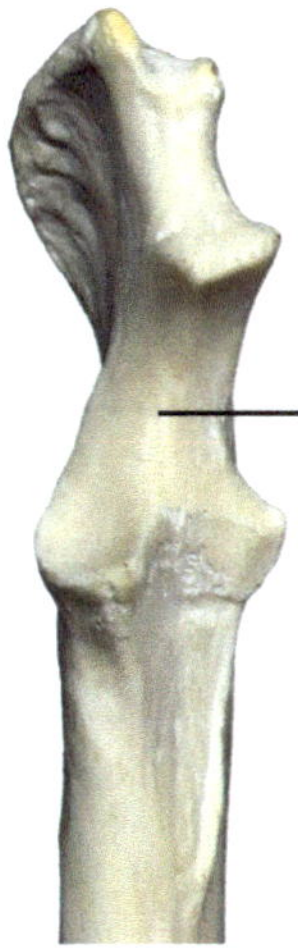

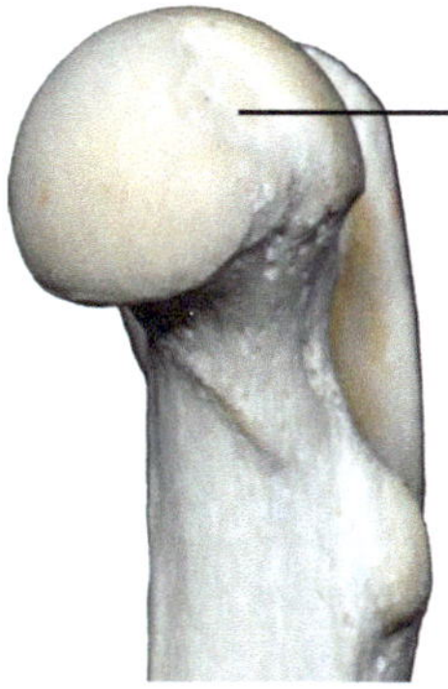

Femur (proximal end)

CHAPTER 3

The Articular and Muscular System

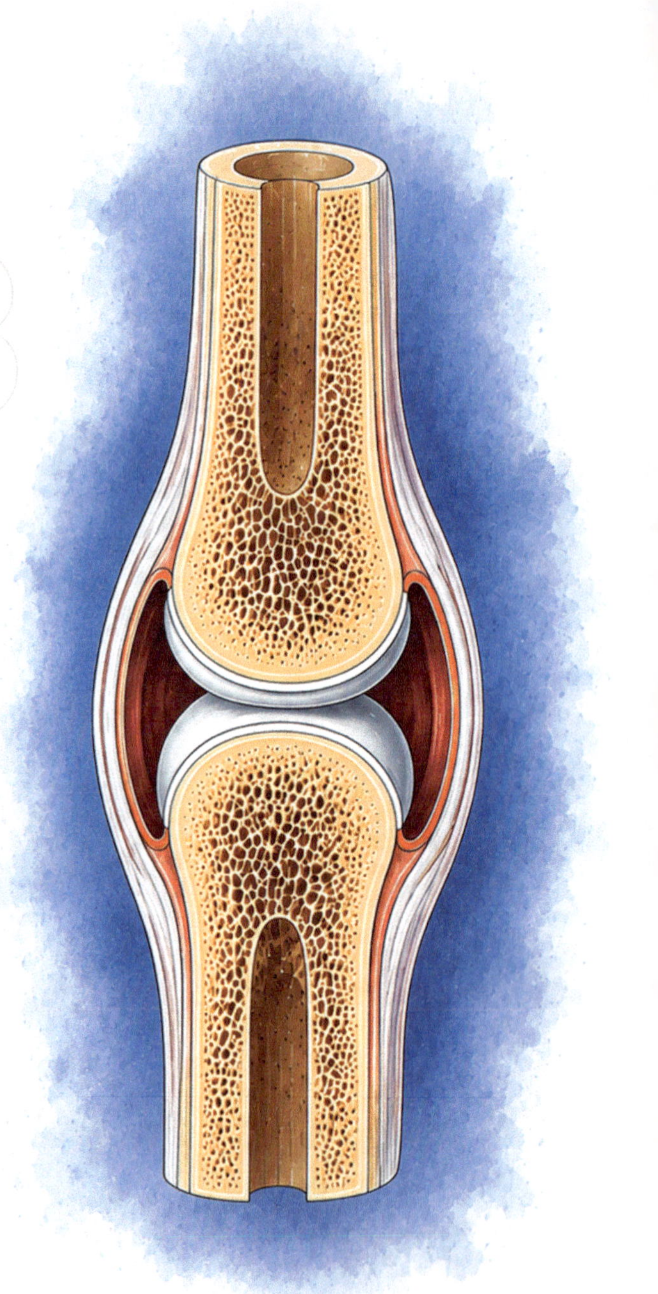

fibrous tissue

very short ⟶ no movement (as in sutures of skul)

Short ⟶ slight movement (as in tibiofibulrar jt.)

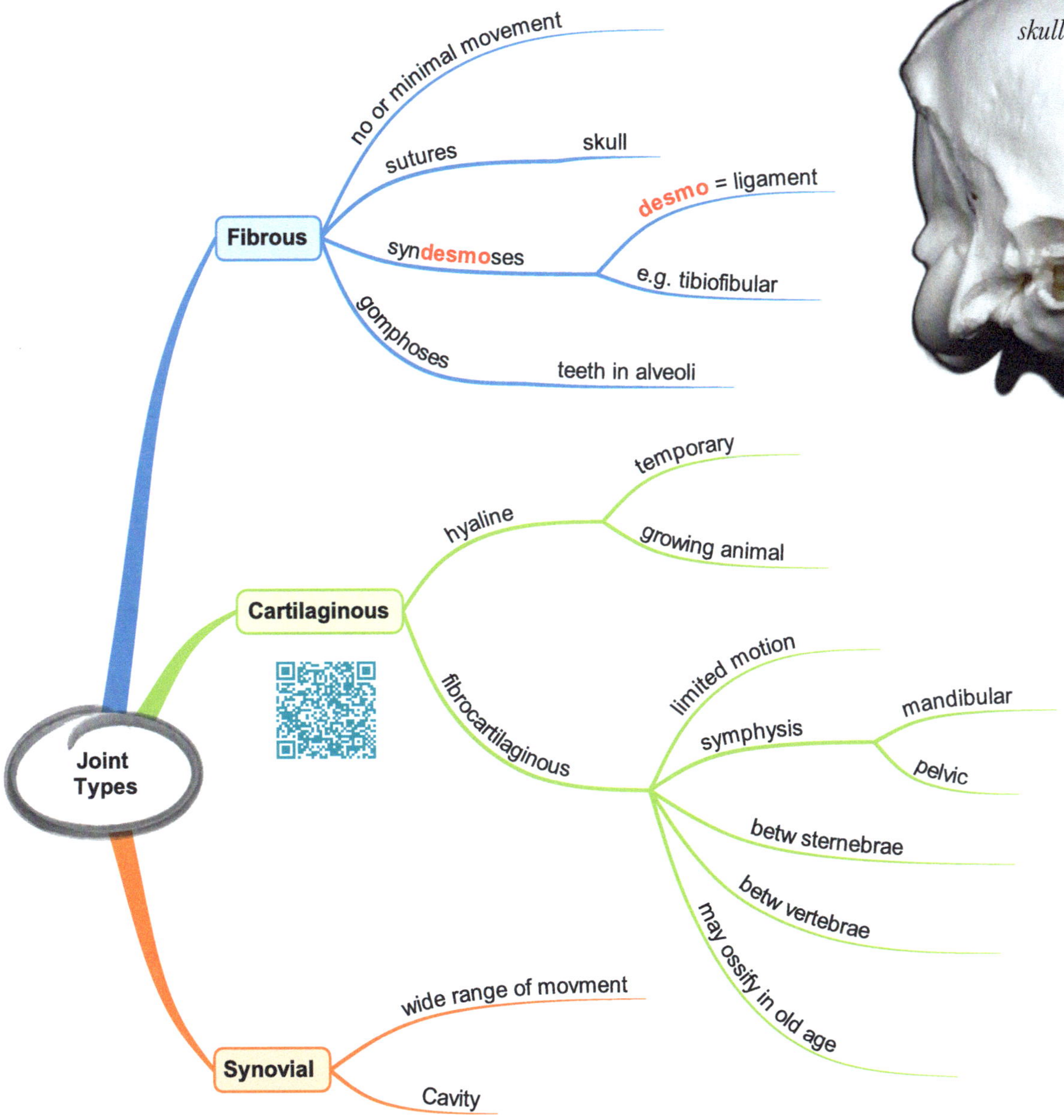

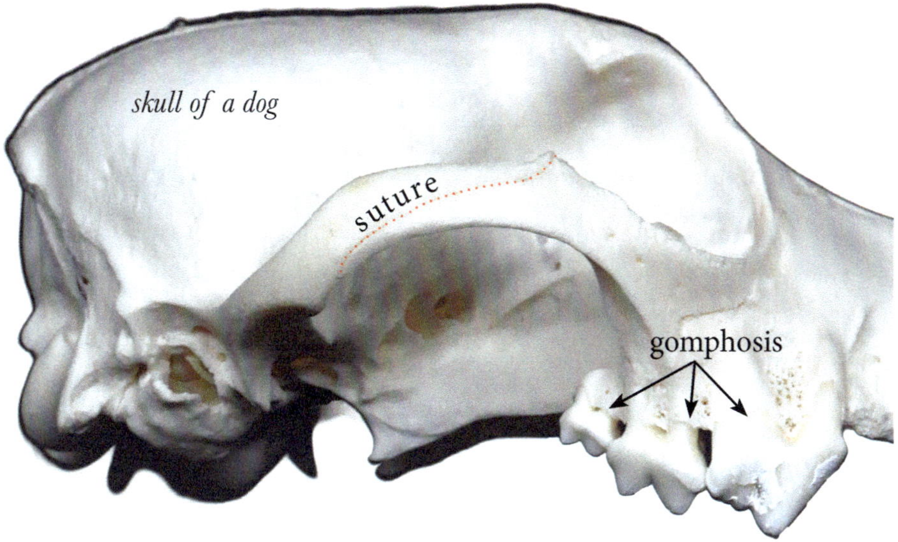

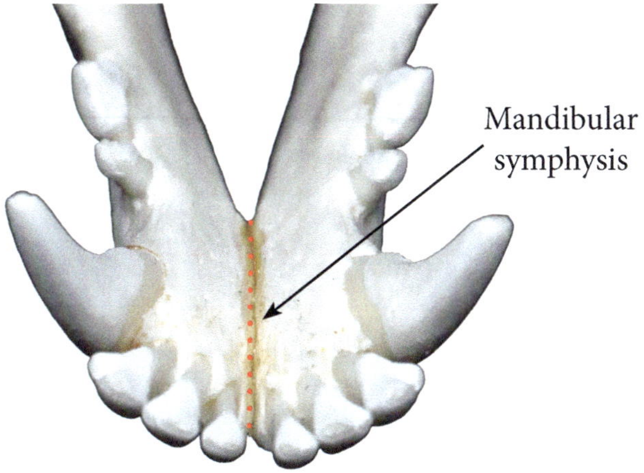

Symphysis jts. are found only in the midline of the body.

Synovial joints

Synovial Jt. Structure

- **Bony articular surface**
- **Articular cartilage** (cover → Bony articular surface)
  - no vessels
  - no nerves
- **Joint**
  - cavity
    - synovial fluid
  - capsule (outer bandaging)
    - sleeve-like
    - fibrous (outer)
    - synovial (inner) — secretes → synovial fluid
- **Ligg.**
  - capsular — thickening of fibrous jt capsule — e.g. shoulder jt.
  - extracapsular — e.g. collateral ligg.
  - intracapsular
    - extrasynovial — *i.e. outside the synovial membrrane*
    - e.g. cruciate ligg
    - e.g. lig. of head of femur
- **Articular discs & menesci**
  - fibrocartilage
  - ingrowth of fibrous capsule — push the synovial membrane (but do not penetrate it)
  - intracapsular (within the jt. cavity) but extrasynovial (are not bathed in synovial fluid)
  - disc — in (tempromandibular jt.)TMJ
  - meniscus — in stifle jt.

SynOvial
=
"Eggs toghether"

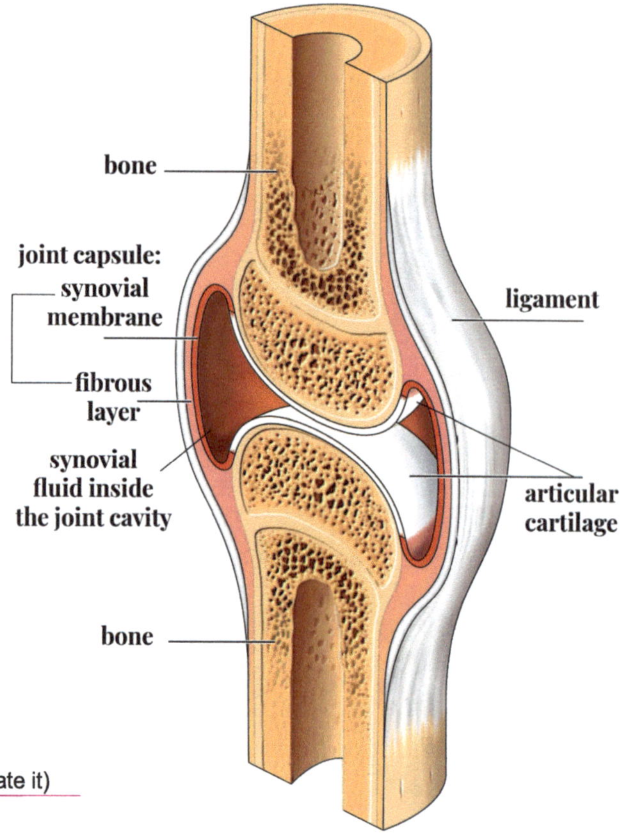

Notice: where the articular cartilage ends the synovial membrane begins.

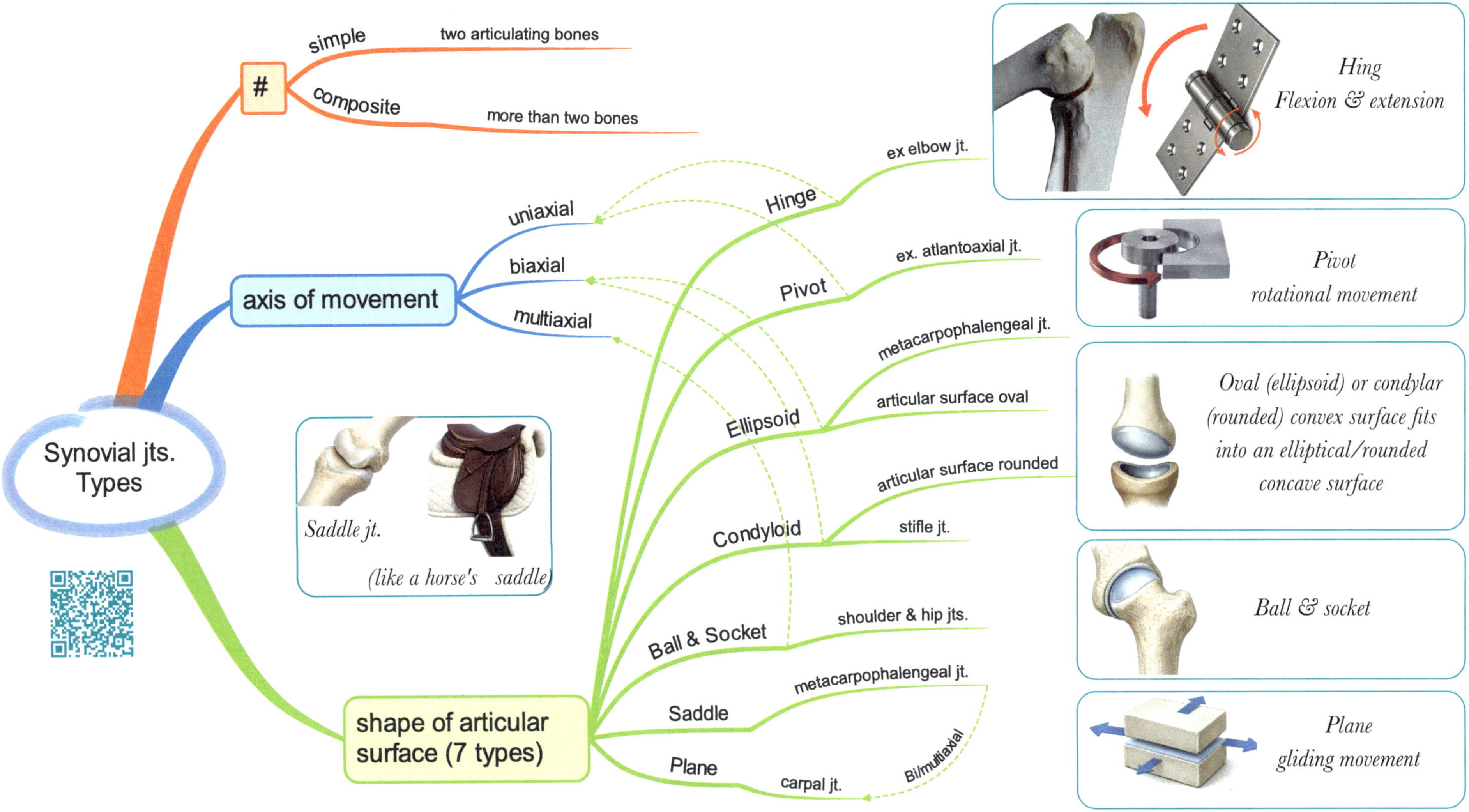

General Rules of Synovial Joints

Hilton's Law: A joint is supplied by the nerves that innervate the muscles moving the joint.

Neurovascular Rule: Major neurovascular bundles lie on the flexor (concave) side of a jt., where they are protected from excessive stretching during movement.

Arterial Rule: Arteries give off articular branches proximal and/or distal to a joint. These branches anastomose to form a periarticular arterial network that ensures collateral blood supply throughout the joint movement.

Naming Muscles
Shape
trapezius
diamond like
Each side (Rt & Lt) is fan shaped, together they are a trapezium.
Rhomboideus
rhomboid-like
deltoideus
delta shaped (triangular)
circule or little orbit
orbicularis oris
orbicularis oculi
sphincter
Relaxed eye open
Contracted eye closed
Location
in the arm (brachium)
brachialis M.
cr. to the tibia
cranial tibial M.
between the ribs (costae)
intercostal Mm
Attachment
sternum
head
sternocephalicus
Position
supraspinatus
above scapular spine
infraspinatus
below scapular spine
popliteus
in popliteal fossa
Action
flexors
flexor carpi radialis
extensors
extensor carpi radialis
supinator
supinates forearm
pronator
e.g. pronator t
Fiber direction
straight
rectus abdominis
transverse
transversus abdominis
oblique
external abdominal oblique
internal abdominal oblique
History related
ex. sartorius M sitting position of tailors in the past

Muscle Rules

- Each muscle has attachments, **Origin** (more stable) & **Insertion** (more mobile).
- In limbs the origin is proximal & the insertion is distal
  - for a muscle to have an action (contract), it should cross at least one joint (except cutaneous Mm.)
- During contraction, the insertion end moves towards the origin. (origin & insertion are relative)
- If a muscle crosses more than one jt. its strong effect will be on the far joint from the origin because the Torque = F x D

- cep = head (part of a muscle that has its own attached tendon).
- 2 heads ---- biceps brachii & biceps femoris
- 3 heads --- triceps
- 4 heads .... quadriceps femoris (not forceps)

Sartor in latin means Tailor. The sartorius muscle got its name from the tailors in the past, used to sit cross-legged while sewing. This position places the thigh in flexion, abduction and external rotation – the same actions performed by the sartorius muscle.

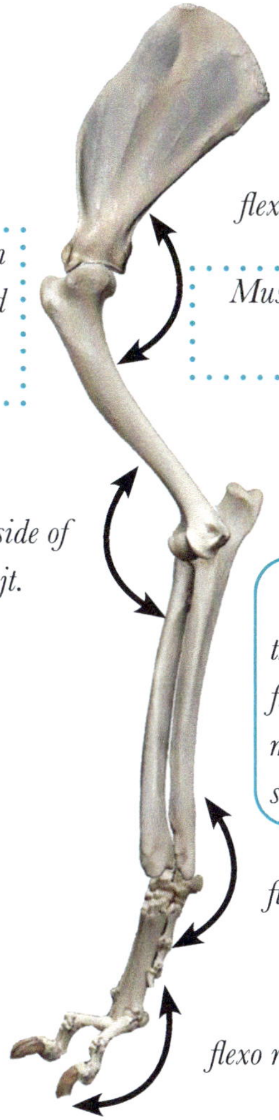

Rule: *muscles crossing the flexor side of a jt. act as flexors of that jt., whereas muscles crossing the opposite side act as extensors.*

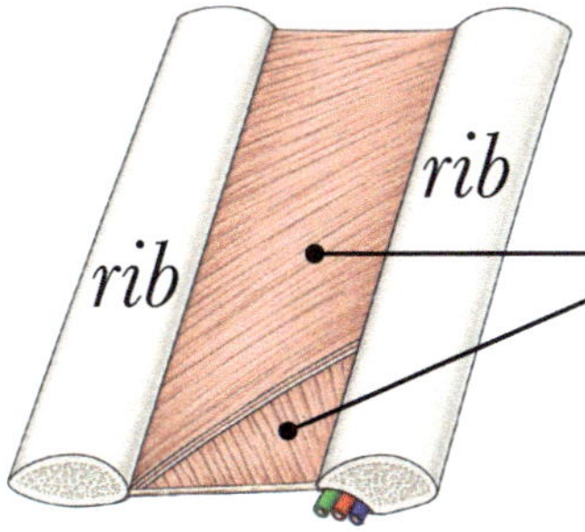

Fascia

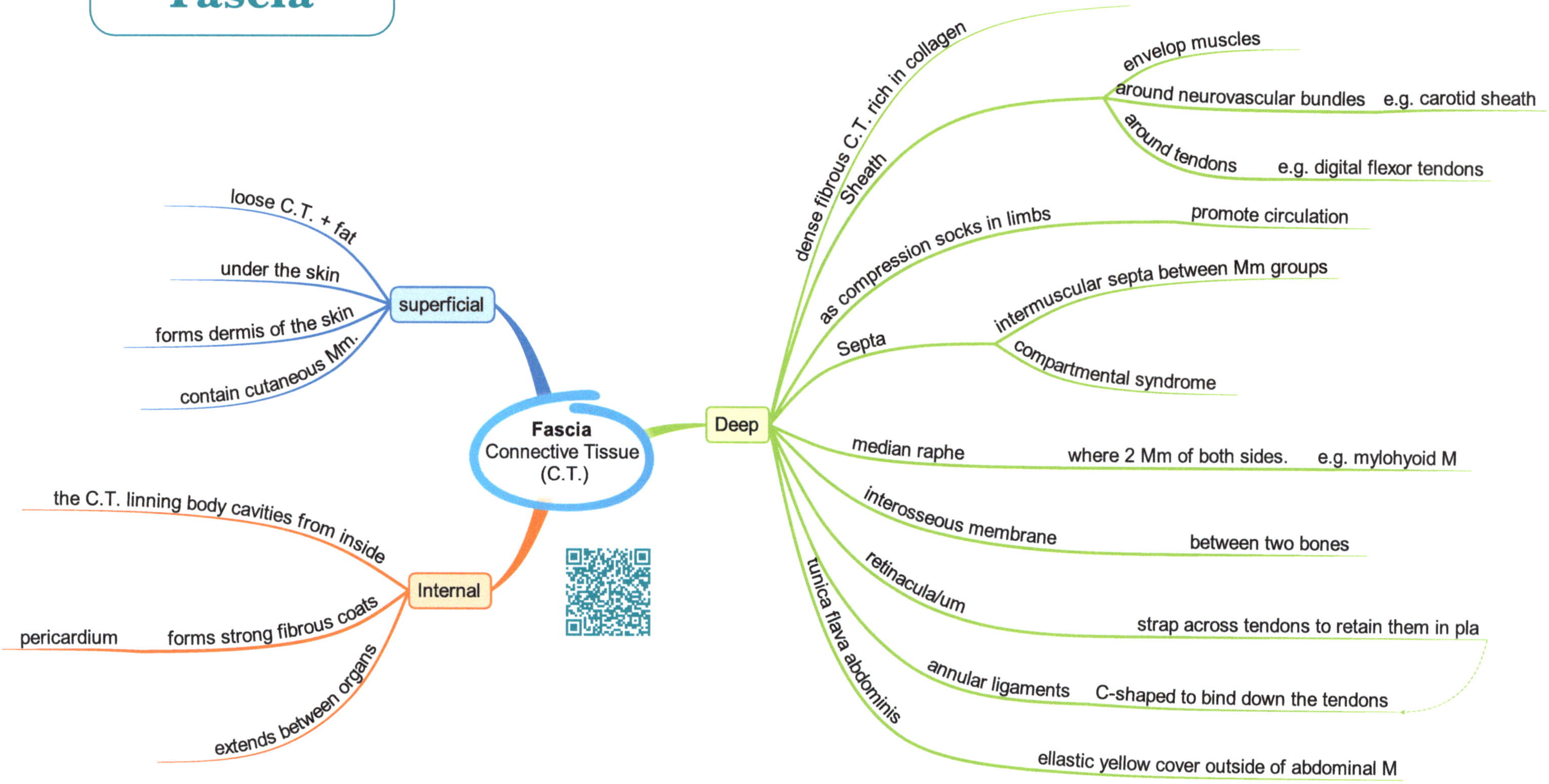

The Connective Tissue (C.T.) forms a continuous, body-wide network, with every C.T. blending seamlessly into the next.

Compartmental syndrome:

*Fascia is tough & inelastic, like a **rigid box**, so it cannot expand.*

When there is a trauma, bleeding, edema, or tight bandages/casts, the contents swell→ Pressure rises because the box cannot exapand → high pressure → Blood vessels compressed → cuts off circulation → Muscle & nerve ischemia → Tissue necrosis (if untreated).

Treatment: ***Emergency fasciotomy*** *(cut the fascia) to release the pressure & restore blood flow.*

Antebrachial fascia:

Caudomedial muscles: enclosed in distinct fascial compartments.

Craniolateral muscles: no distinct fascial compartments.

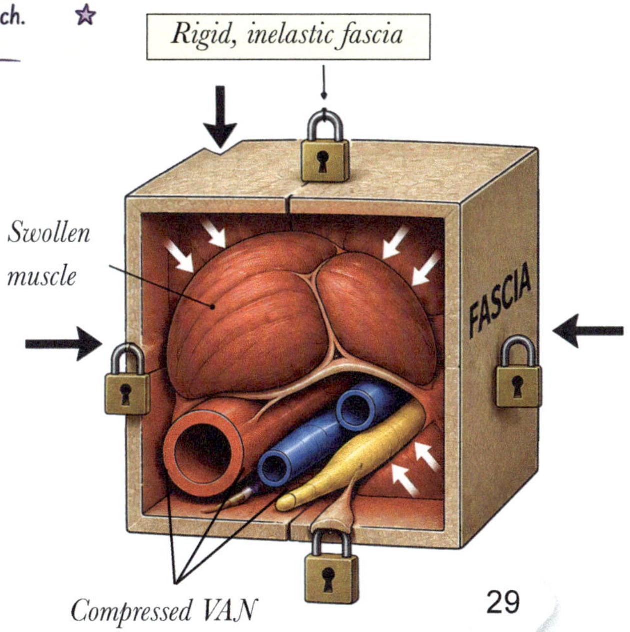

Tendons & Aponeurosis *both are fibrous tissue attaching muscles to bones.*

Tendons are rounded (cord or cable like), Aponeuroses are flat.

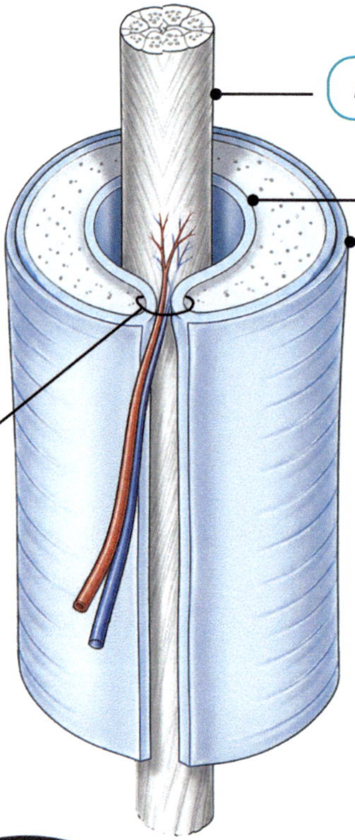

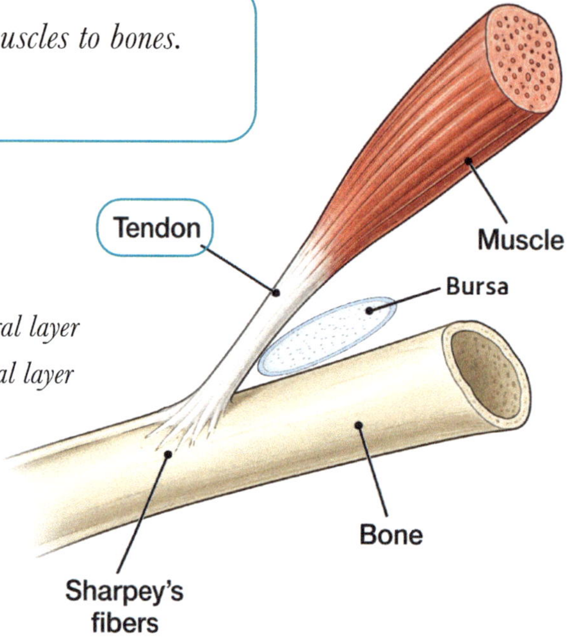

Bursa:

- *pillow like, closed sac filled with synovial fluid.*
- *inflamation → bursitis*
- *e.g. inflamation of infraspinous bursa → shoulder lameness.*
- *e.g. inflammation of olecranon bursa → hygroma & elbow lameness.*

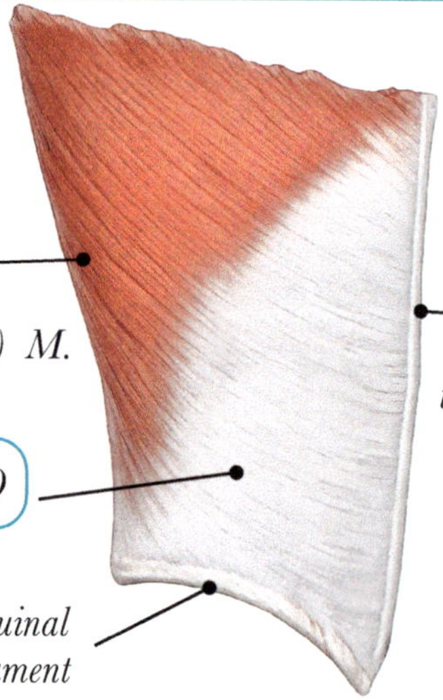

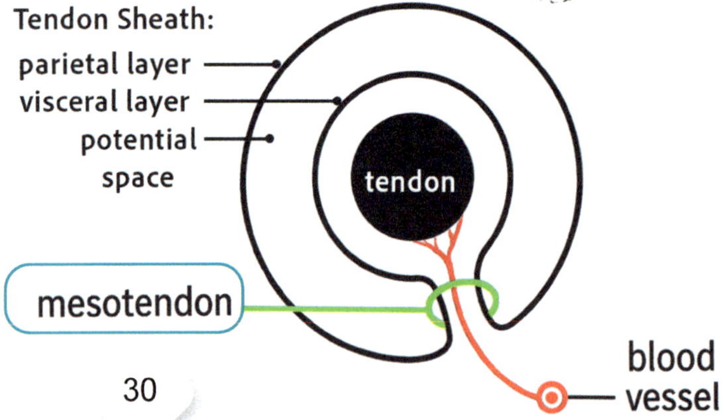

The muscle is enveloped by C.T called Perimysium which will continue to ensheath the tendon (peritendineum). On reaching the bone, the peritendinum blends with the periosteum while the collagen fibers penetrate the cortex of the bone forming the perforating fibers of Sharpey.

From Principles to Practice

You have now mastered the foundational language of anatomy—the vocabulary, the structural organization of the skeleton, the classification of joints, and the principles of muscular nomenclature. In other words, you have learned the grammar of anatomy.

In the chapters that follow, we move from basic concepts to regional anatomy. We will apply this anatomical grammar to the forelimb, exploring it not as a collection of separate structures, but as an integrated and highly functional system.

As you progress through the forelimb, continue to rely on the principles and structural rules introduced in the previous chapters. They will serve as your roadmap for understanding the relationships between bones, joints, muscles, vessels, and nerves.

Now, let us begin our journey into the anatomy of the forelimb.

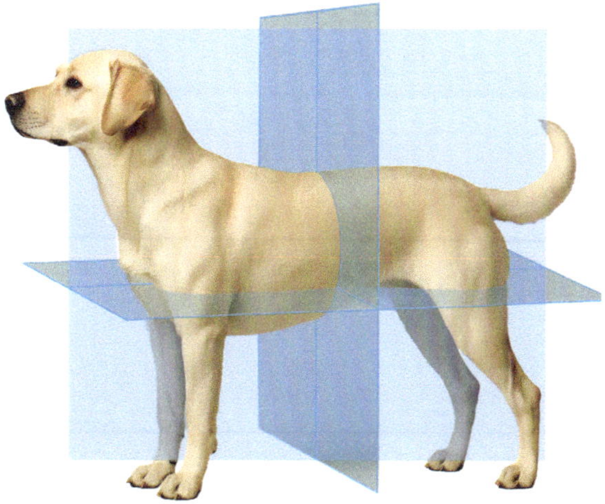

Quick Recall Check

Before turning the page, take a moment to review:

Can you name the four major types of bones you will encounter in the forelimb?
(Flat, long, short, and sesamoid bones)
Are you comfortable with the directional terms that guide anatomical description?
(Dorsal, palmar, plantar, cranial, caudal, proximal, and distal)

You will use these terms repeatedly as we explore the muscles, blood vessels, nerves, and other structures of the forelimb.

CHAPTER 4

Osteology of the Forelimb

REGIONAL BONES: • SCAPULA • HUMERUS • RADIUS/ULNA • MANUS

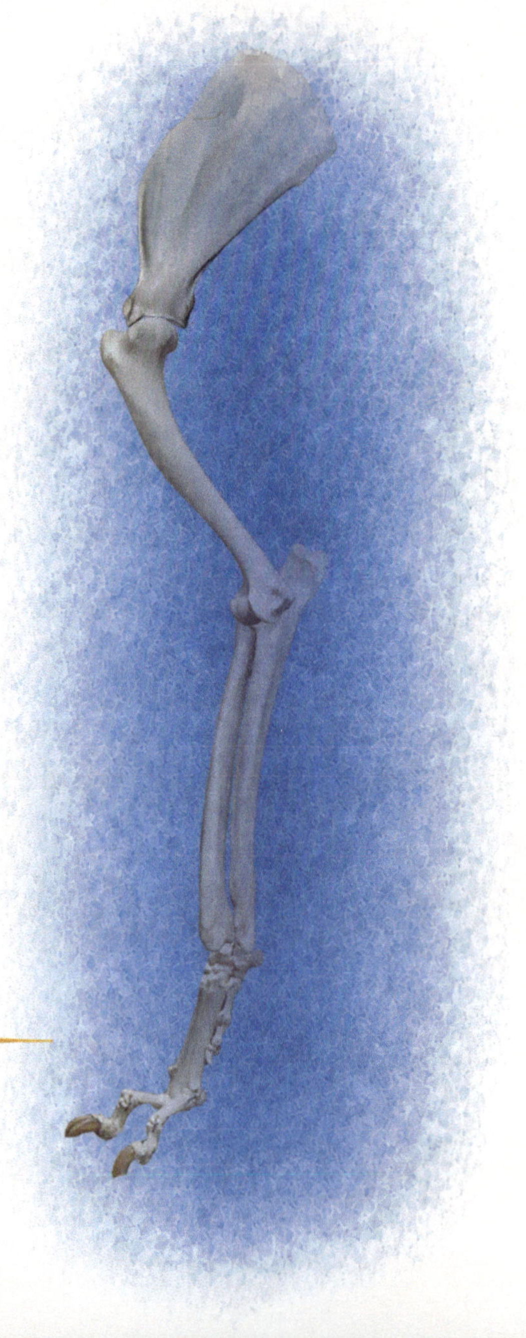

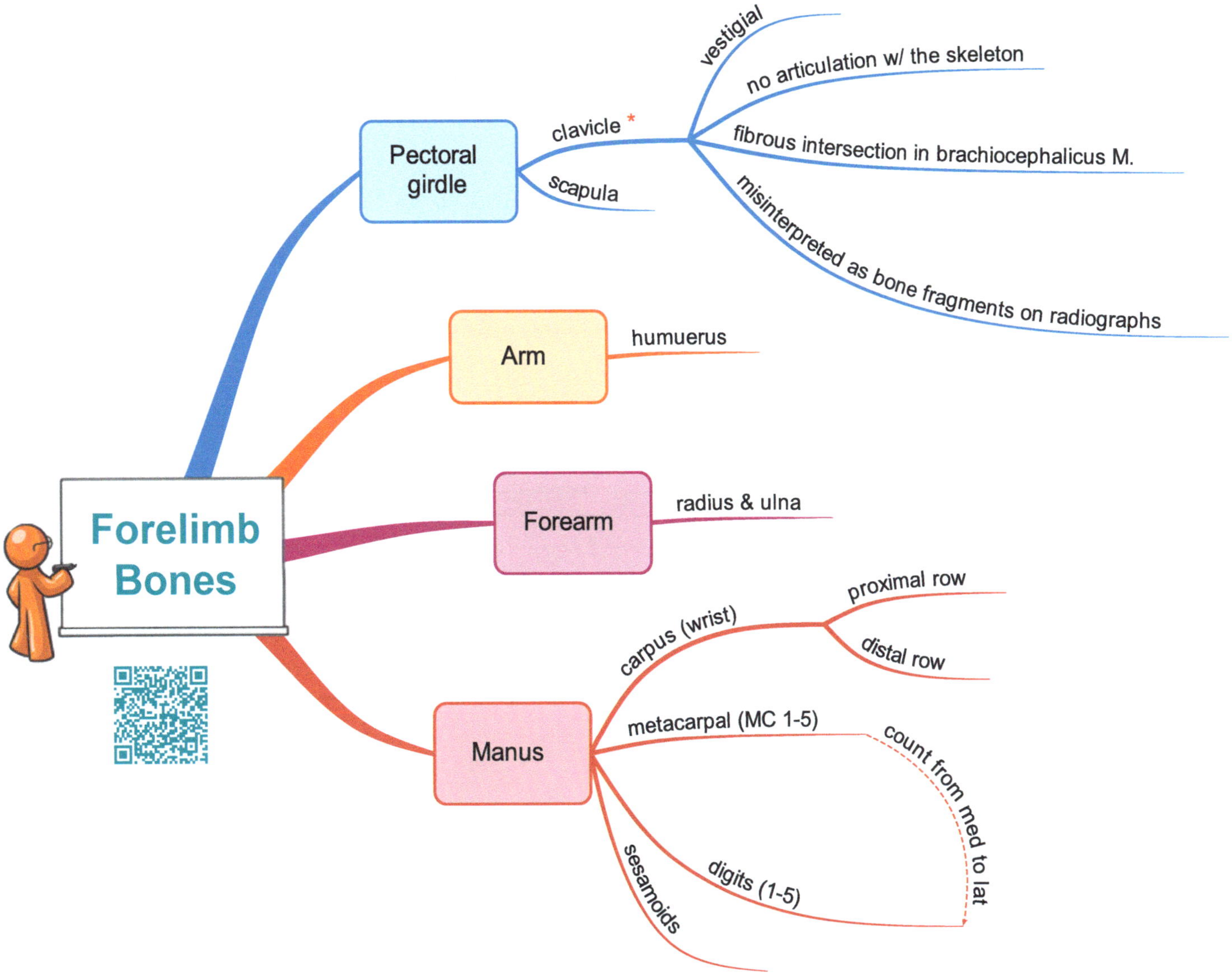

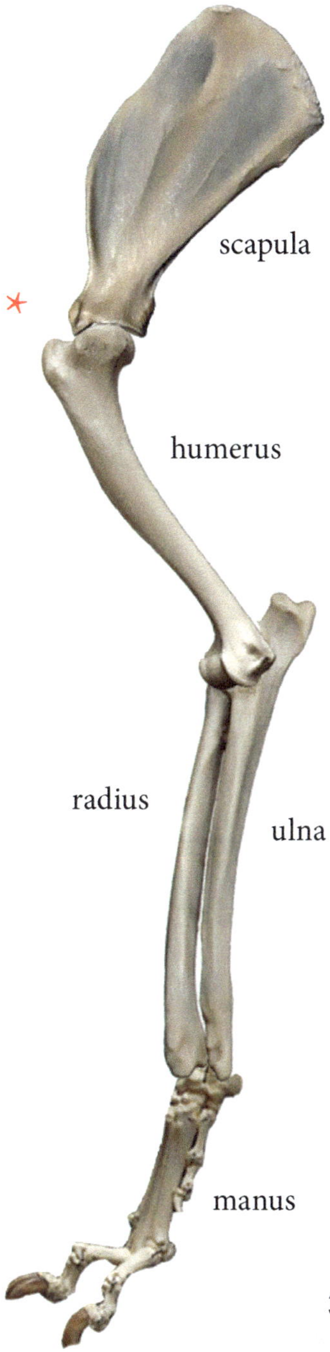

The dog possesses a vestigeal clavicle at the tendineous intersection of the brachiocephalicus M. The cat has a small clavicle that can be mistaken for a chip fracture on radiographs.

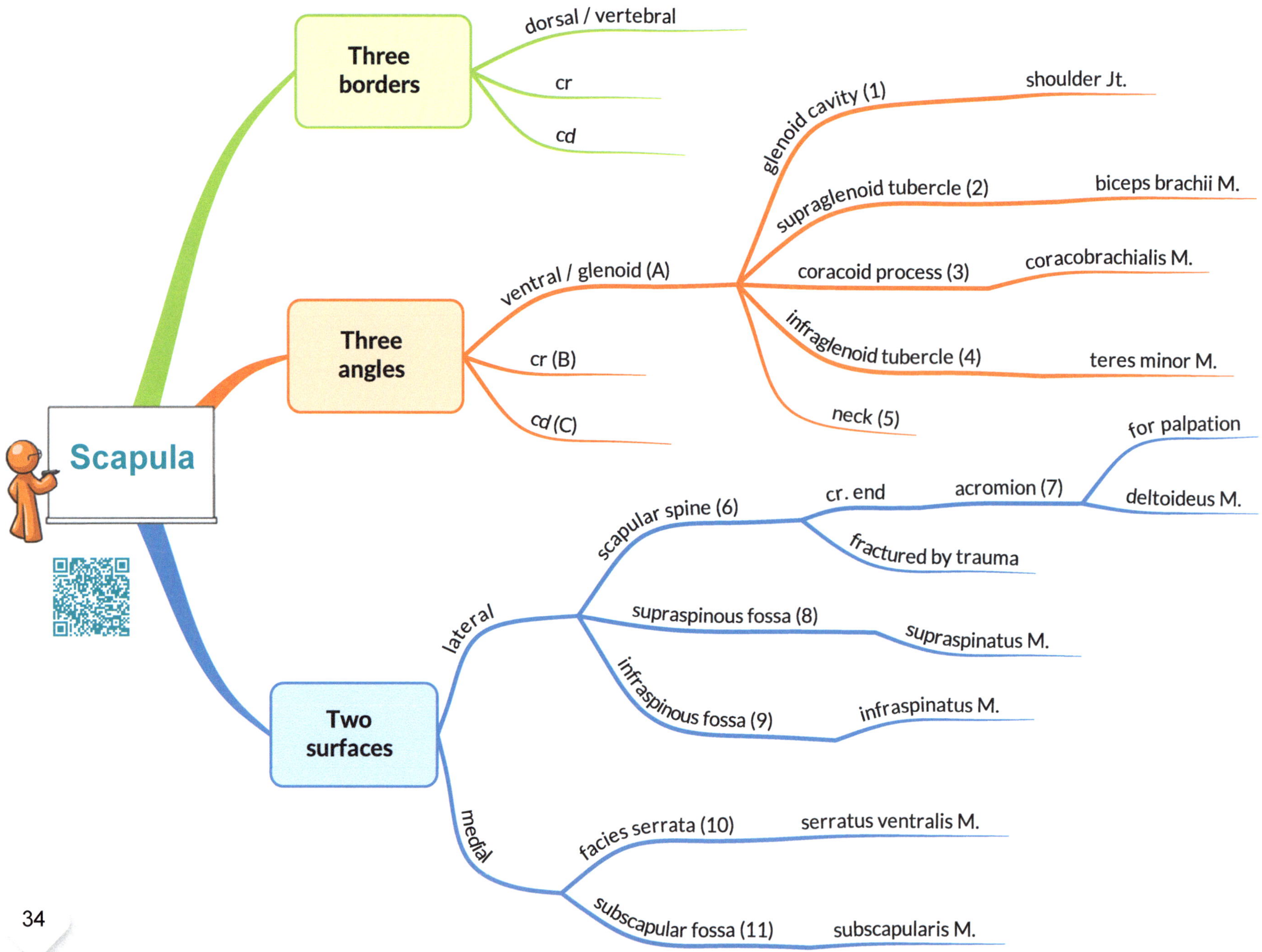

Scapula
Three borders
dorsal / vertebral
cr
cd
Three angles
ventral / glenoid (A)
glenoid cavity (1)
shoulder Jt.
supraglenoid tubercle (2)
biceps brachii M.
coracoid process (3)
coracobrachialis M.
infraglenoid tubercle (4)
teres minor M.
neck (5)
cr (B)
cd (C)
Two surfaces
lateral
scapular spine (6)
cr. end
acromion (7)
for palpation
deltoideus M.
fractured by trauma
supraspinous fossa (8)
supraspinatus M.
infraspinous fossa (9)
infraspinatus M.
medial
facies serrata (10)
serratus ventralis M.
subscapular fossa (11)
subscapularis M.

Left Scapula

Lateral surface

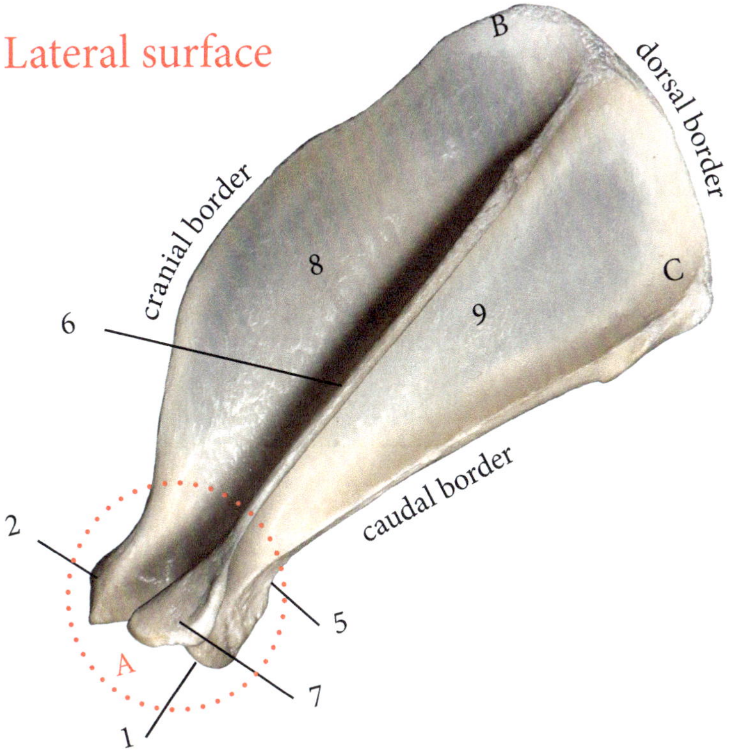

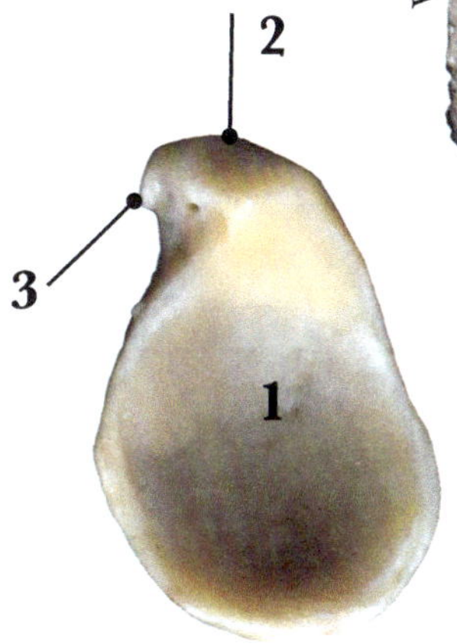

Ventral angle

Medial surface

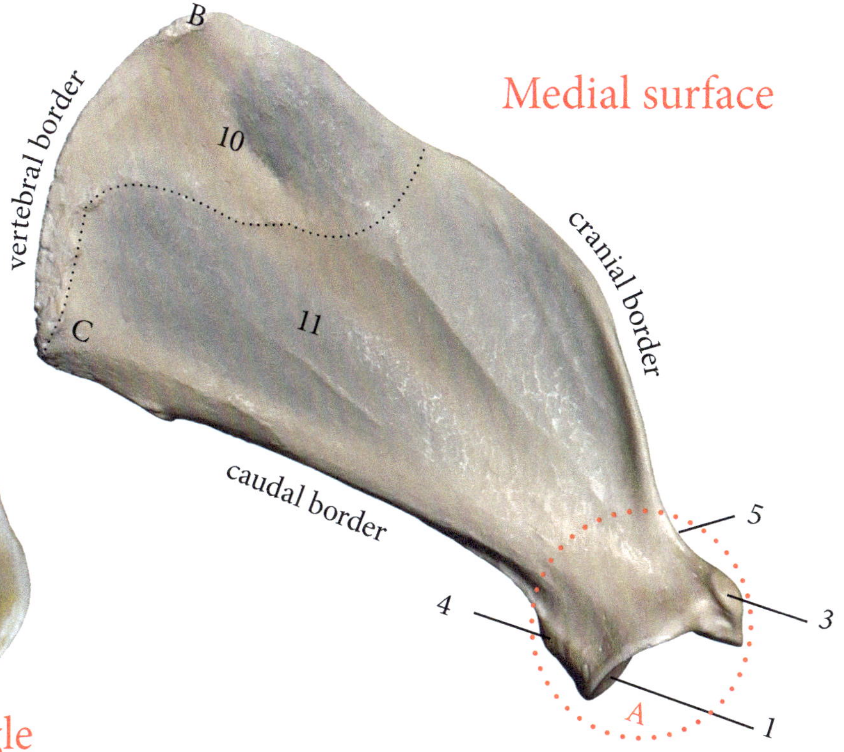

Suprascapular n. passes deep to the acromion and should be avoided in acromial osteotomy.

Fracture of scapula is rare as it is surrounded by muscles.
Fracture of the spine due to trauma.
Fracture of the neck can damage the suprascapular n.
Avulsion fracture of acromion due to strong pull of deltoideus can damage the suprascapular n.
Avulsion fracture of supraglenoid tubercle due to strong pull of biceps brachii M.

Humerus

- Prox
  - head (1)
    - caudal
    - art w/ scapula
    - neck (2)
    - jt. mice
  - greater tubercle (3)
    - lateral
    - palpable
    - site for
      - bone marrow sampling
      - cancellous bone grafting
      - Isupra & infraspinatus Mm
  - intertubercular gro (4)
    - biceps tendon passage
  - lesser tubercle (5)
    - medial
    - subscapularis M.
- Shaft
  - deltoid tuberosity (6)
    - lat
    - deltoideus M.
  - tricipital line (7)
    - lat
    - triceps brachii M.
    - tuberosity for teres minor (8)
  - brachialis /musculospiral gro (10)
    - passage of brachialis M.
  - teres major tub (9)
    - med
    - teres major & latissimus dorsi Mm.
- Dist / condyle
  - capitulum (14)
    - lat
    - art w/ ulna
  - trochlea (15)
    - med
    - arti w/ radius & ulna
  - two fossae
    - olecranon (18) (long word)
      - cd
      - deep
      - receives anconeal process
    - radial (19) (short word)
      - cr
      - small
      - receives radial head
  - foramen
    - supratrochlear (16)
      - dog
      - nothing passes through
    - supracondylar (17)
      - cat
      - brachial a.
      - median n.
  - epicondyles
    - med (11)
      - flexor of carpus & digit
    - lat (12)
      - extensors of carpus & digit
      - lat epicondylar crest (13)

The Captin *(capitulum)* is Late *(lateral)*

The tricepital line extends between two tuberosities; deltoid & teres minor

Surgeons find it harder to reach the humerus because it is surrounded by muscles & spiraly twisted.

Left humerus

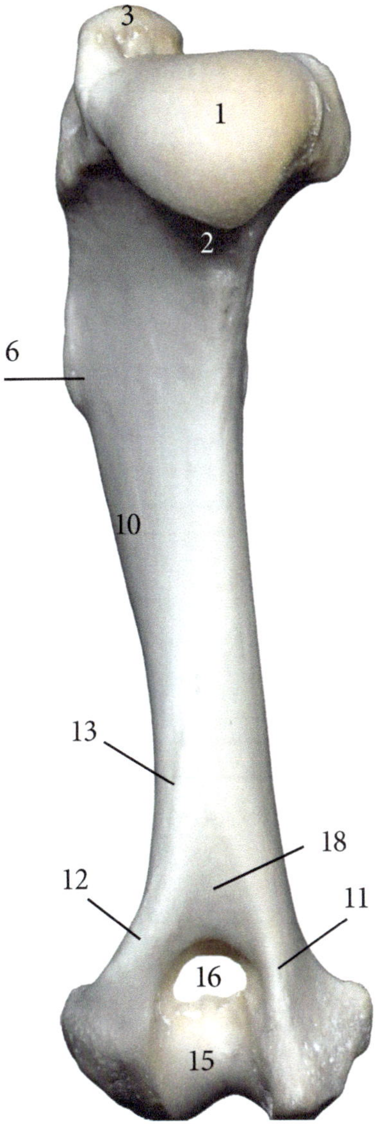

Caudal view

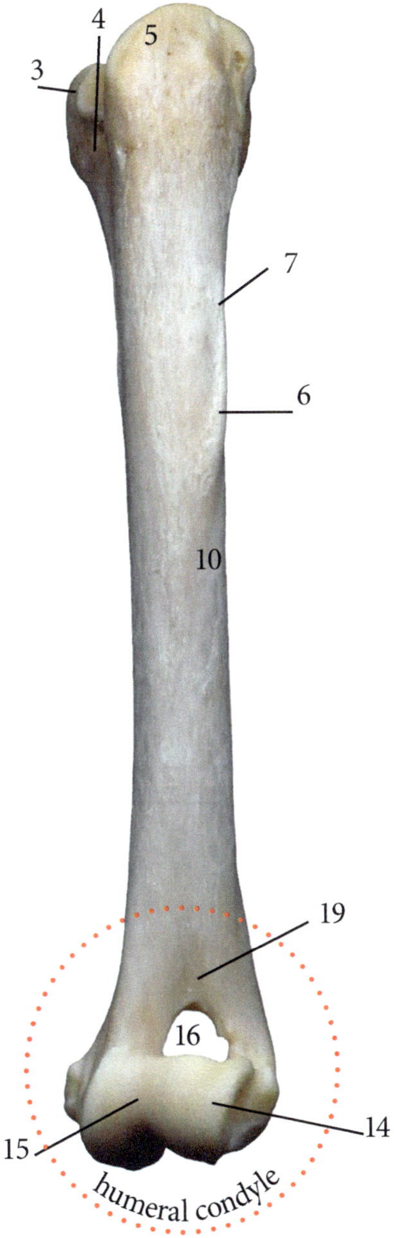

Cranial view

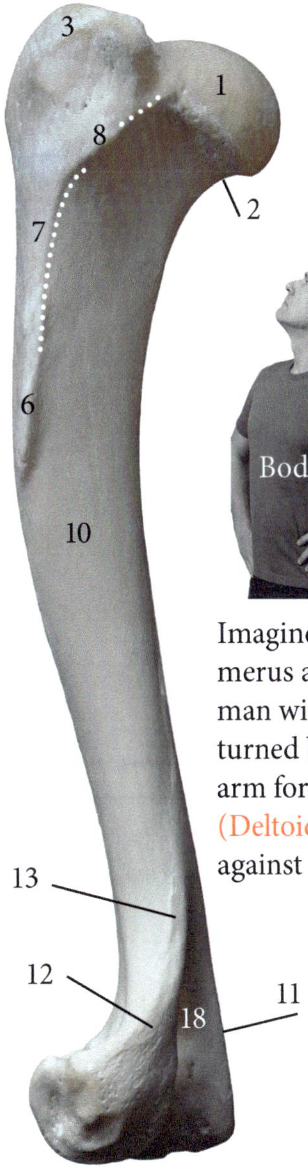

Lateral view

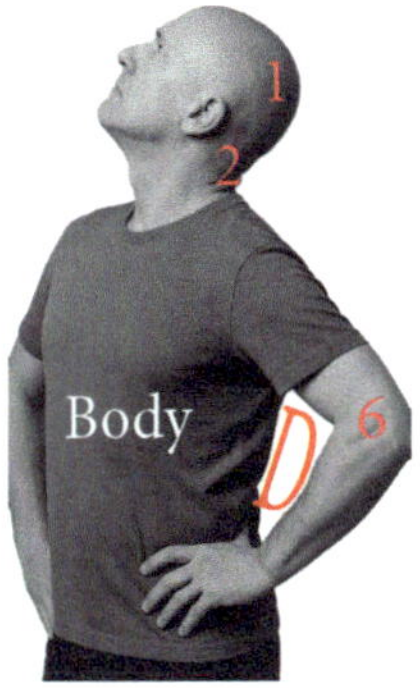

Imagine the humerus as a bald man with his head turned back & his arm forming a D (Deltoid tuberosity) against his waist.

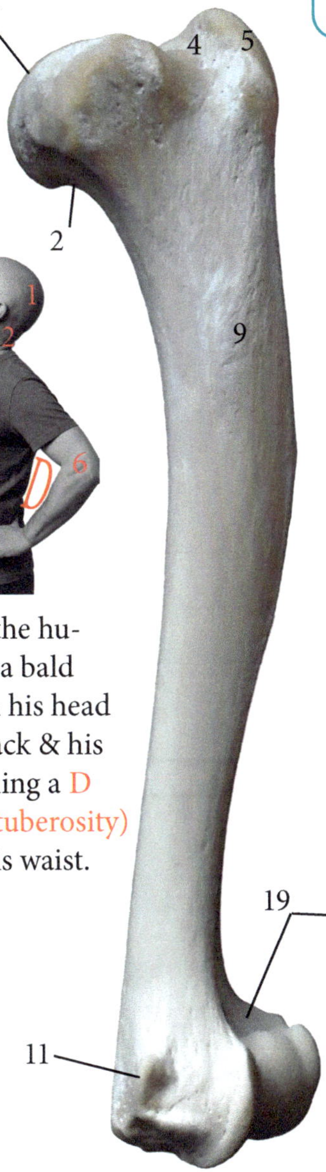

Medial view

The humerus is the strongest long bone in the body. The femur is the second.

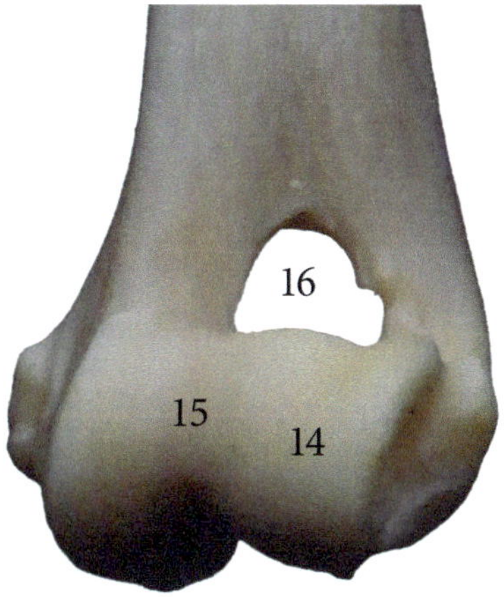

distal extremity humerus (dog), left, cr. view

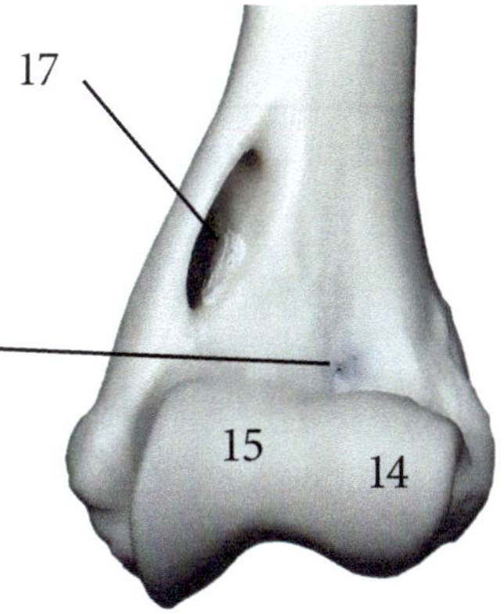

distal extremity, left humerus (cat), cr. view

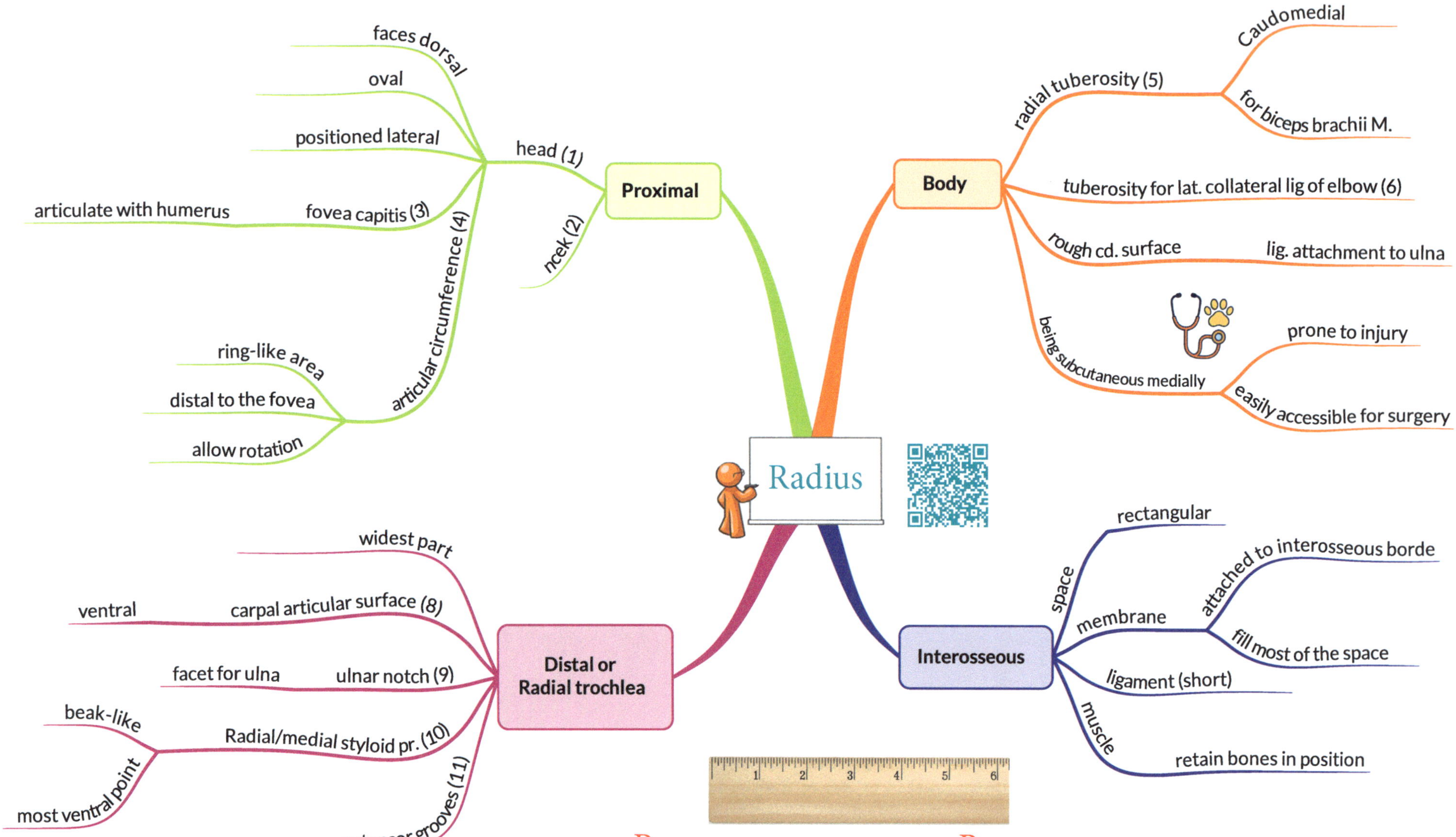

- The Dog's Radius is flat craniocaudally like a Ruler
- The Radius Rotates around the ulna. The ulna is relatively fixed, so muscles that rotate the forearm (pronation & supination) insert on Radius.
- The radial tuberosity (on radius) & ulnar tuberosity (on ulna) are located caudomedial.

Notice

The head of radius is proximal

The head of ulna is distal

Ulna head = Under (distal)

Radius head = Roof (proximal)

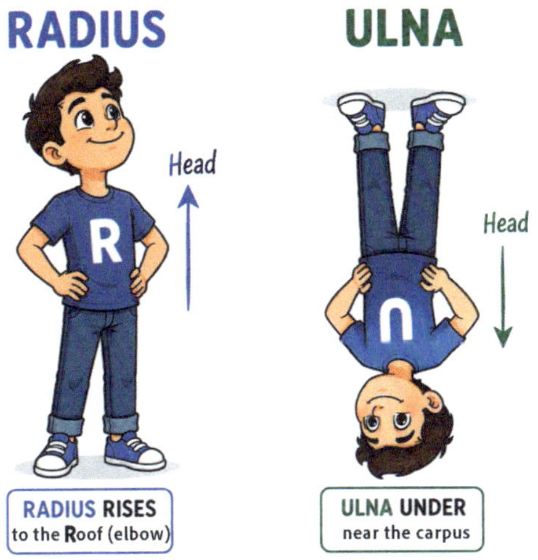

Poor healing of distal radial fractures is due to poor blood supply

because the the nutreint a of radius enters the radius hight @ the junction of prox & middle thirds & extend distally . In addition to a br. of cd interosseous A., so the distal extremity is poorly vascularized.

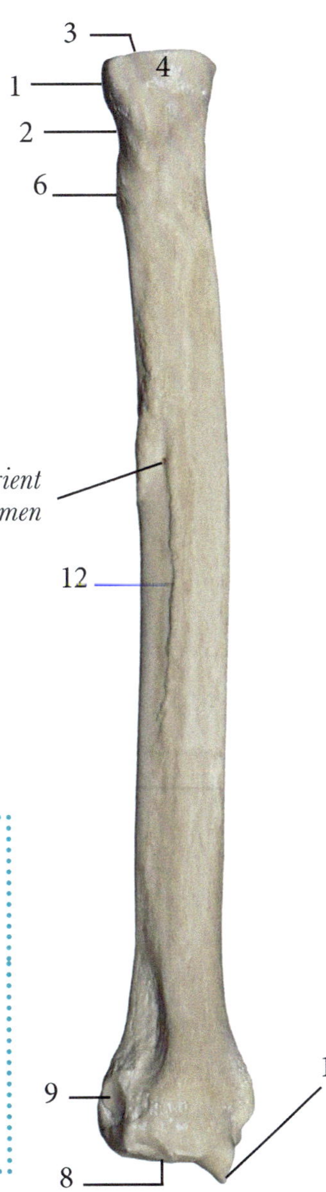

Caudal view

Cranial view

Radius & Ulna are fixed in Pronation by the prox. interosseous ligament.

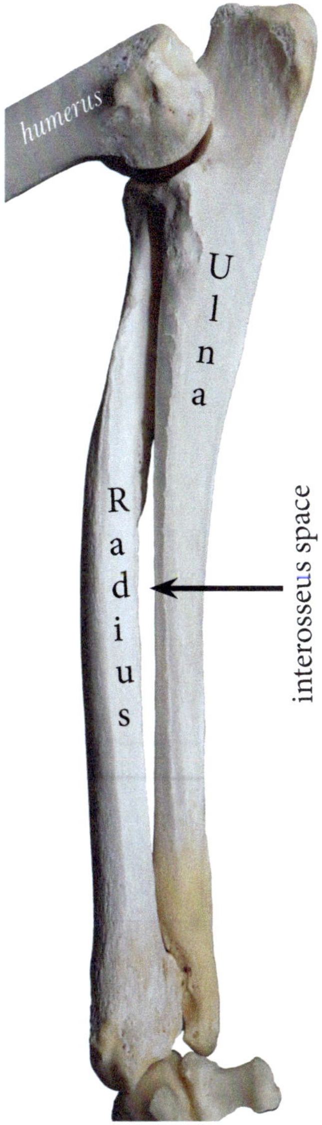

Ulna

- **Proximal Extremity**
- **Body**
  - Ulnar tuberosity (7)
    - caudomedial
    - distal to the **med** coronoid pr.
    - biceps & brachialis insertion
  - obliquely crosses the radius caudally
  - interosseous border (8)
    - lig. attachment to radius
- **Head/Distal**
  - positioned laterally directly under the skin
  - articular circumference (9)
    - w/ ulnar notch of radius
  - lateral styloid process (10)

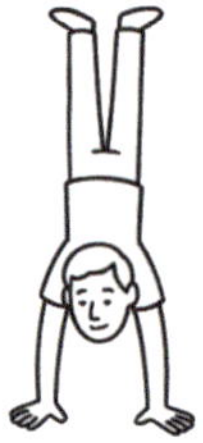

Distal ulna: exposed laterally under skin, with no muscle cover, making it prone to injury.

Stylus = pen
Styloid = pen like

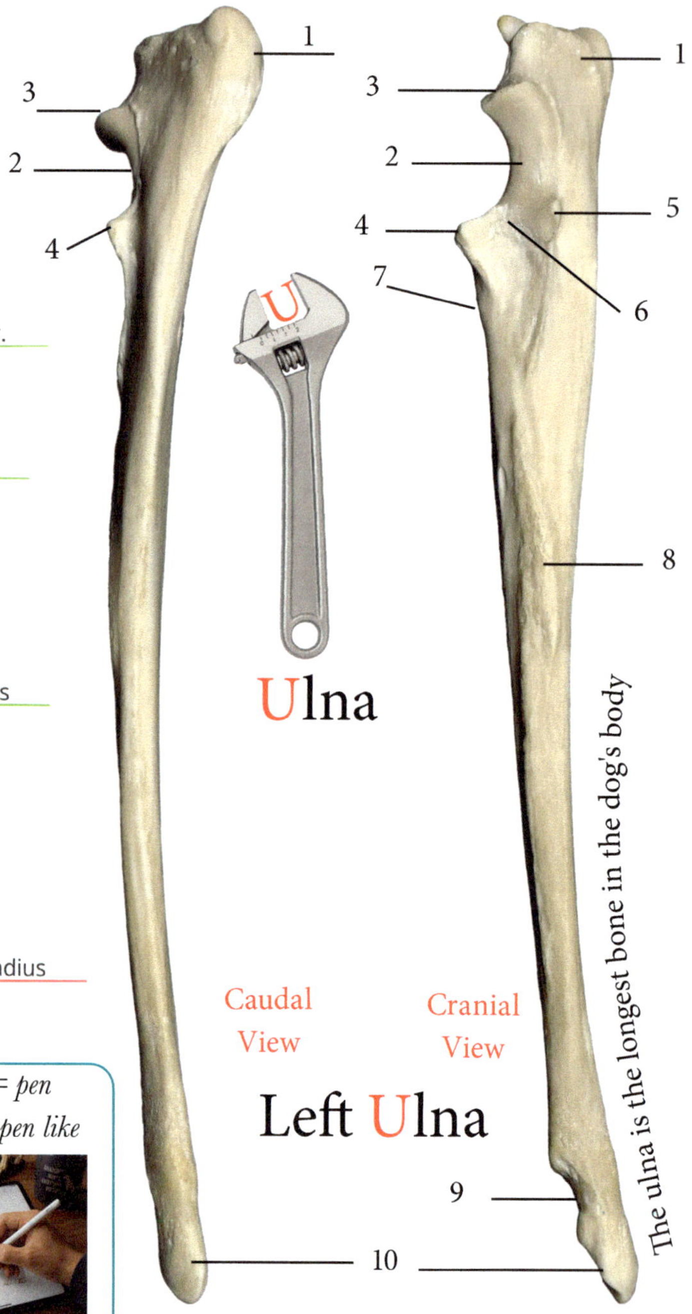

Ulna, prox. extremity (Olecranon)

- concave medially
- positioned medially @ the elbow
- olecranon tuber (1)
  - point of elbow
  - insertion of elbow extensors
- Trochlear notch (2)
  - hinge jt. w/ humeral trochlea
  - cranial
  - Cresent-shaped
  - ends
    - proximally by
      - Anconeal pr. (3)
        - beak-like (A-shaped)
        - fuses @ 5-6 M.
        - Ununited Anconeal Process (UAP)
        - in extension sits in olecranon fossa
    - distally by
      - Two processes & a notch in between
      - coronoid processes
        - medial (larger) (4)
        - lateral (5)
      - radial notch (6)
        - between coronoid processes
        - articulate w/ radius articular circumference

The ANCONEAL PROCESS prevents side & rotational movements of the extended elbow → stability. If fractured or UAP → arthrosis & lameness of of elbow jt.

(FCP) Fragmented medial & lateral Coronoid Processes.

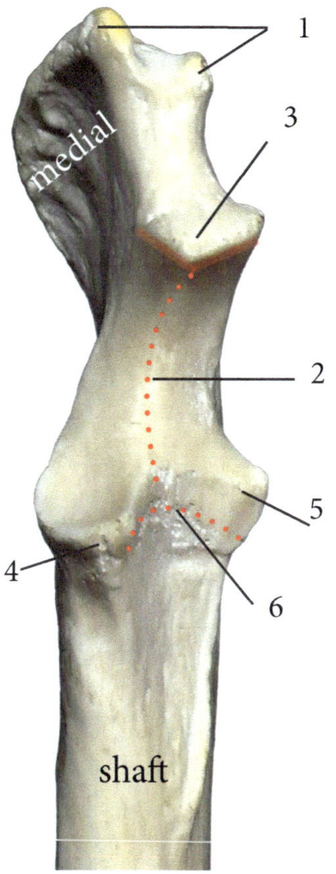

Left Ulna

RADIUS & ULNA

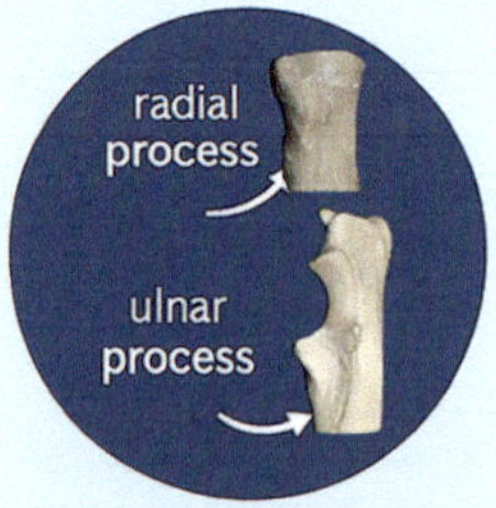

TUBEROSITIES

both have a process on the medial side

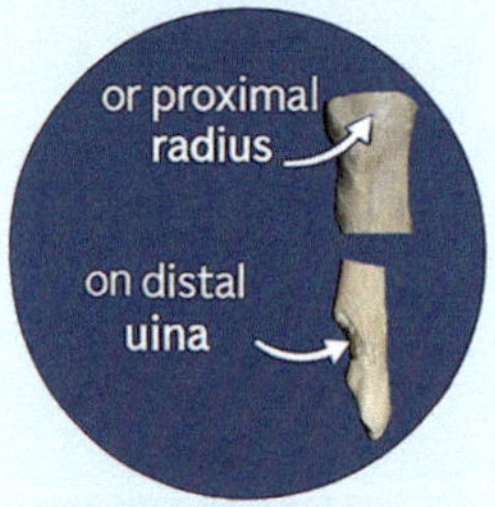

ARTICULAR CIRCUMFERENCE

on proximal radius & distal ulna

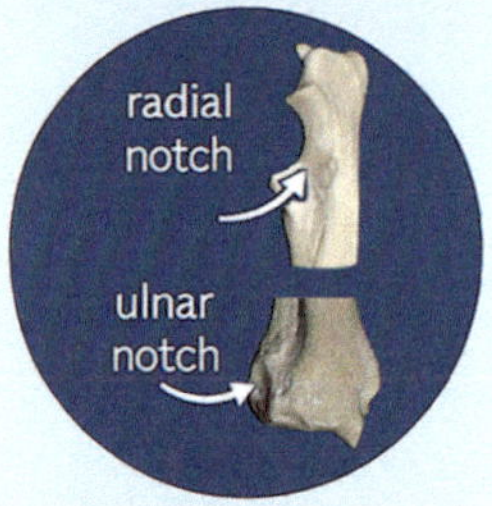

NOTCH

radial notch on ulna and ulnar notch on radius

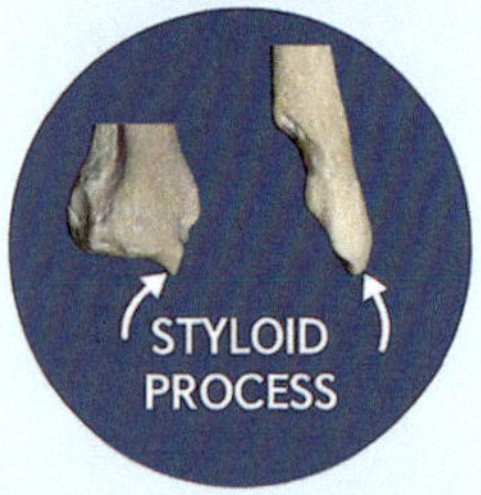

STYLOID PROCESS

both have a styloid process at the distal end

The olecranon subcutaneous bursa can get inflaammed (hygroma)

Avulsion fracture of the olecranon due to the pull of triceps brachii M.

Fracture of olecranon will be directed cranioproximally by the traction force of the triceps brachii M.

The anconeal process stabilizes the extended elbow by resisting lateral & rotational movements.

Ulna: has a notch proximally (radial notch for radial articular circumference).
Radius: has a notch distally (ulnar notch for ulnar art. circumference).

Remember: The Radius is Regular (i.e. head is proximal & the notch distal).
The Ulna is Upside down (i.e. head distal and the notch proximal).

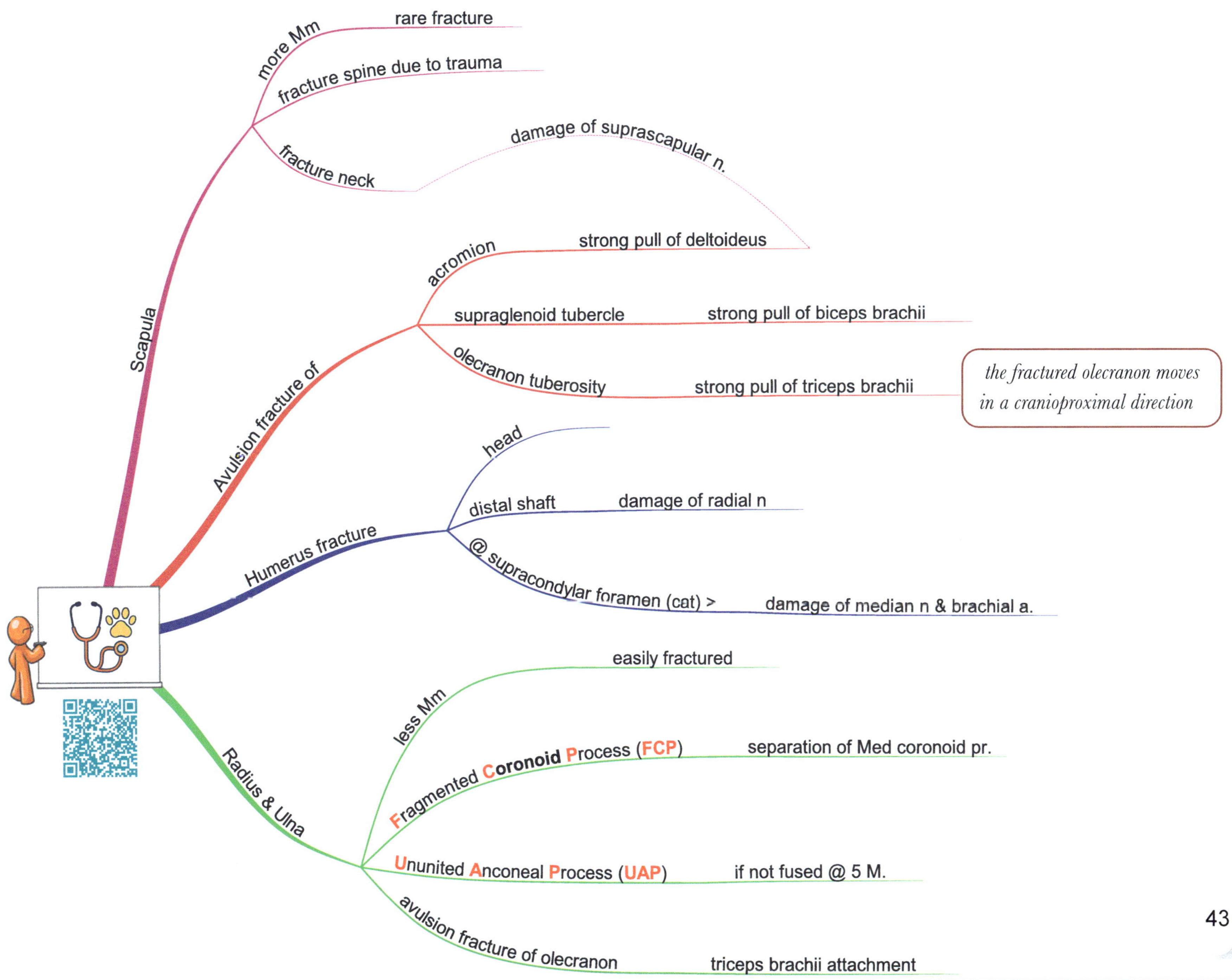
Scapula
more Mm
rare fracture
fracture spine due to trauma
fracture neck
damage of suprascapular n.
Avulsion fracture of
acromion
strong pull of deltoideus
supraglenoid tubercle
strong pull of biceps brachii
olecranon tuberosity
strong pull of triceps brachii
the fractured olecranon moves in a cranioproximal direction
Humerus fracture
head
distal shaft
damage of radial n
@ supracondylar foramen (cat) >
damage of median n & brachial a.
Radius & Ulna
less Mm
easily fractured
Fragmented Coronoid Process (FCP)
separation of Med coronoid pr.
Ununited Anconeal Process (UAP)
if not fused @ 5 M.
avulsion fracture of olecranon
triceps brachii attachment

Manus

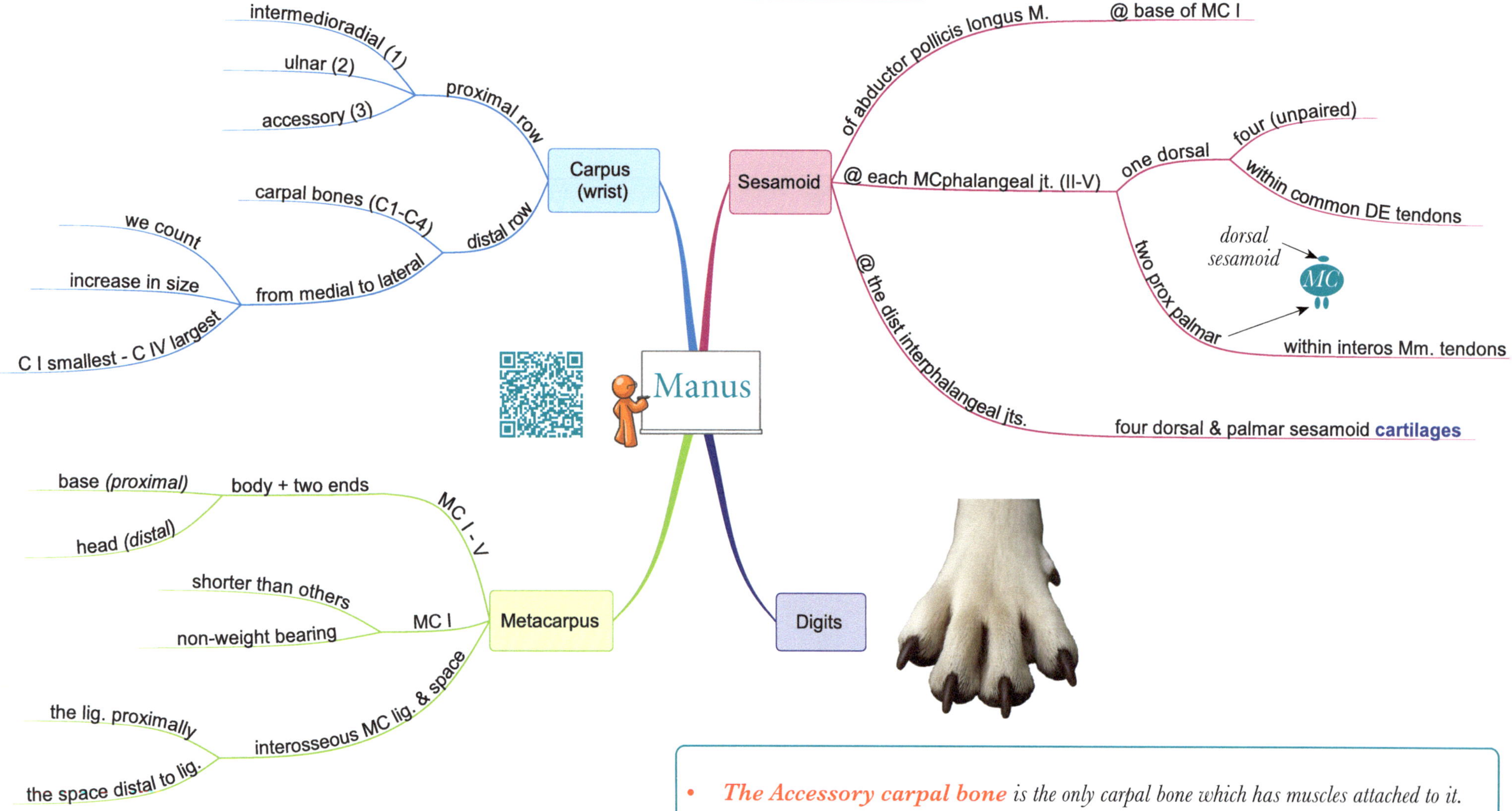

- ***The Accessory carpal bone*** *is the only carpal bone which has muscles attached to it.*
- *Two muscles insert on the accessory carpal bone; the ulnaris lateralis and the flexor carpi ulnaris (both its humeral & ulnar heads).*

Manus

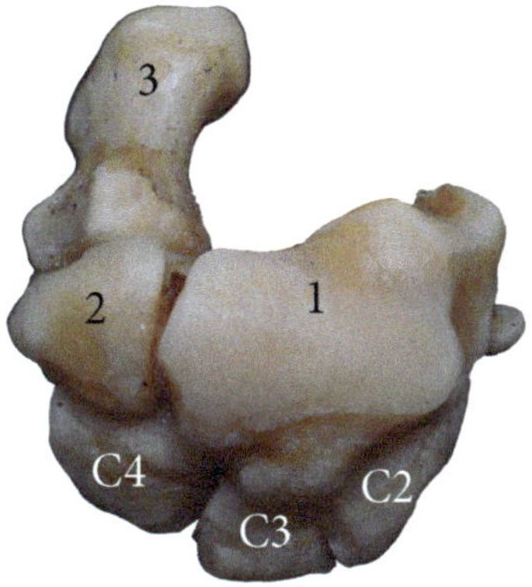

Right carpal bones
Craniodorsal view

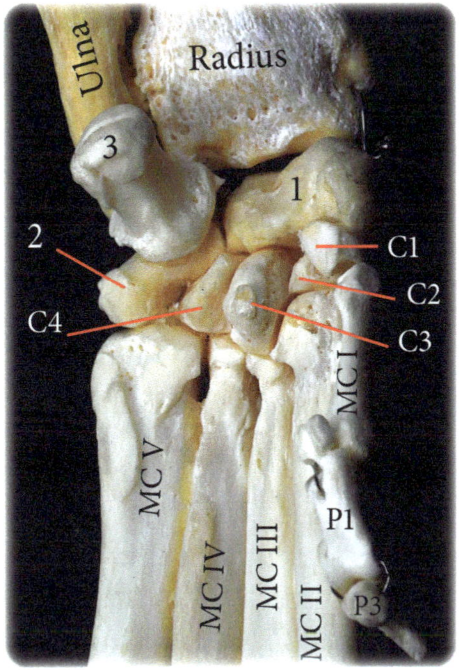

Left manus, palmar view

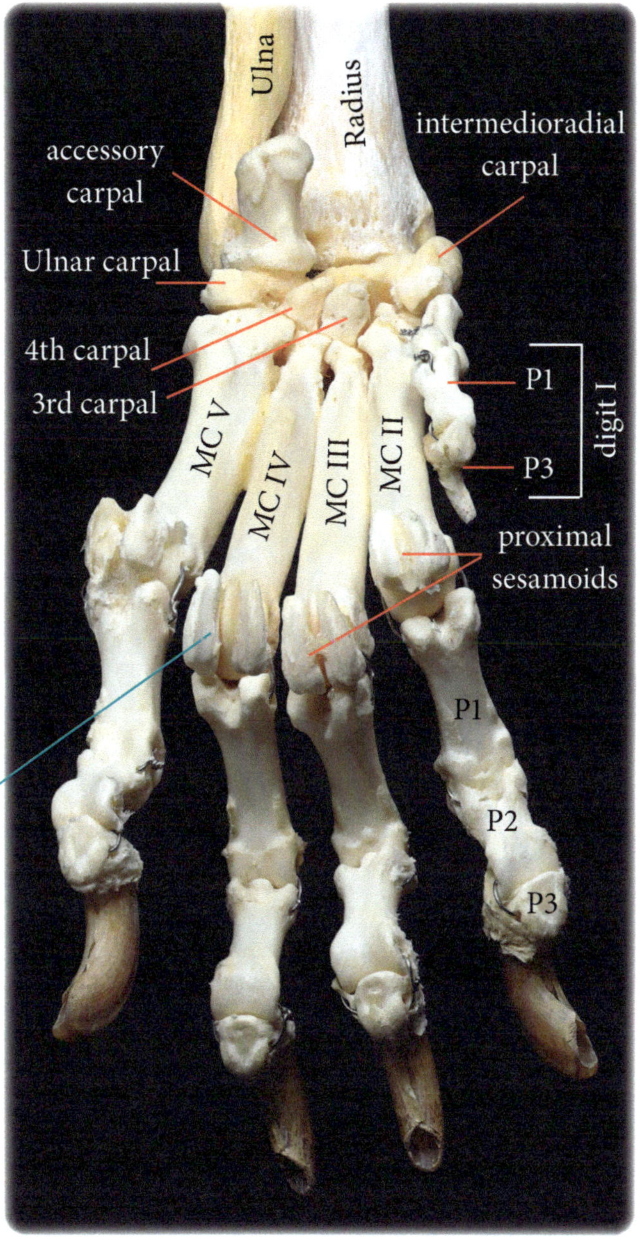

Cross section of Metacarpus at MCP jt.

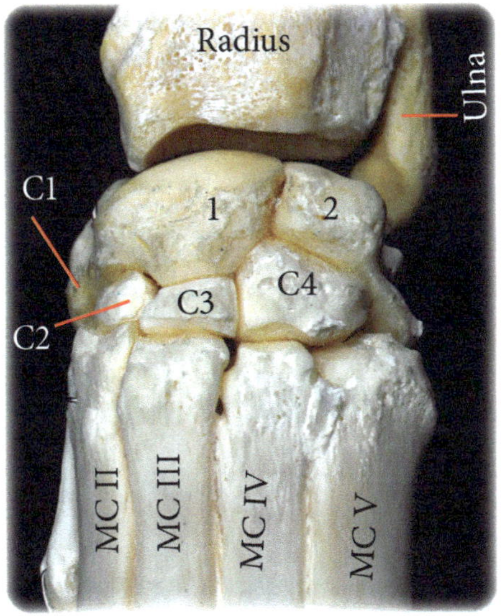

Left manus, dorsal view

Left manus, palmar view

Digits "toes" I-V

- **Digit I**
  - rudemintary
  - two phalanges (no P2)
    - P1
    - P3
  - "the dewclaw"
- **Digits II-V** (three phalanges)
  - P1 (prox phalanx)
  - P2 (mid phalanx)
  - P3 (dist phalanx)
- **P1 & P2**
  - like MC
    - base (proximal)
    - body
    - head (distal)
- **P3**
  - ungual crest (1)
    - bony "shelf" surrounds P3
    - separated from P3 by a space
    - the space filled with bed of the claw
  - ext pr (2)
    - common DE tendon
  - ungual pr (3)
    - claw-shaped
    - dist ext of P3
    - covered by the claw itself
  - flexor tubercle (4)
    - palmar
    - for deep DF tendon

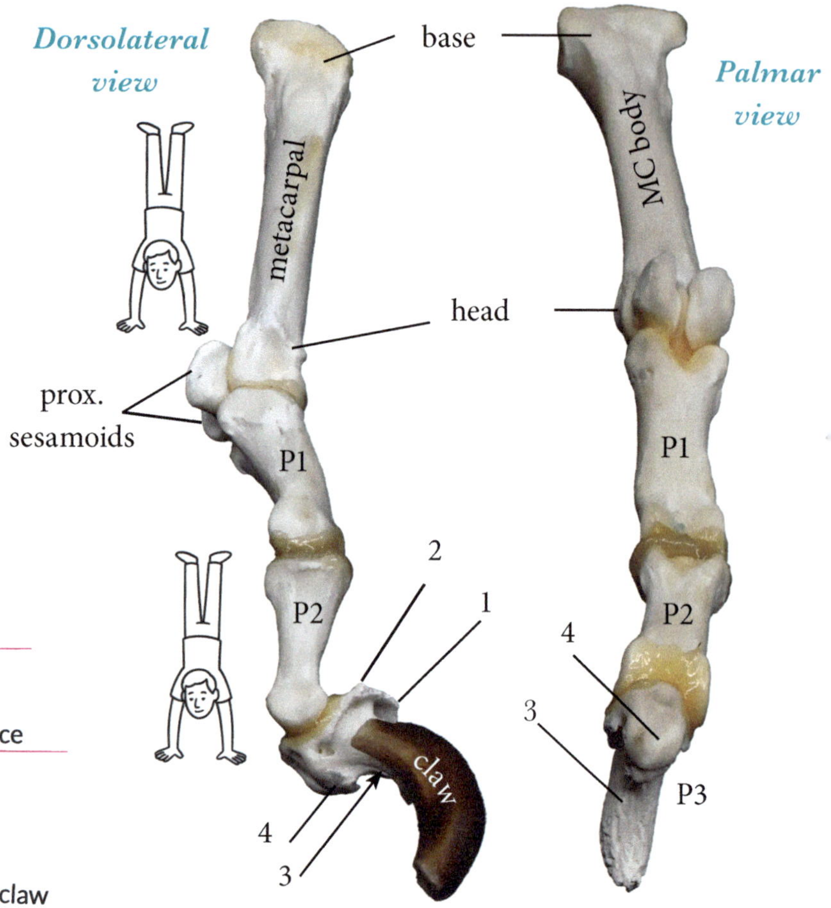

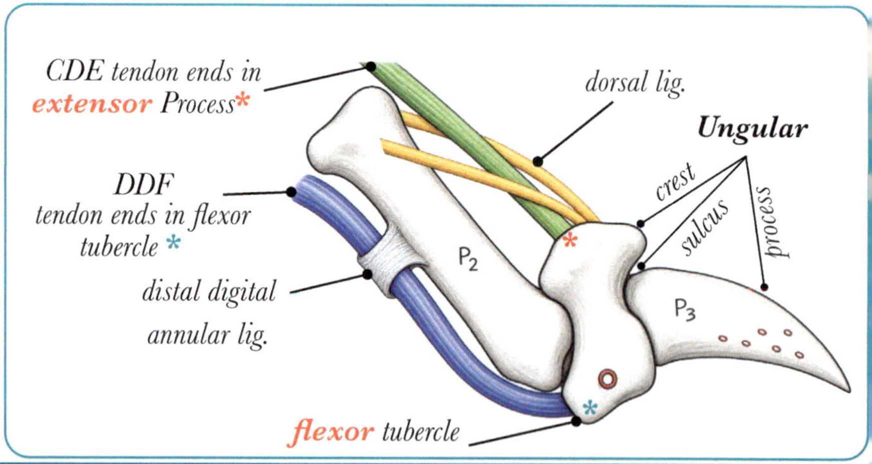

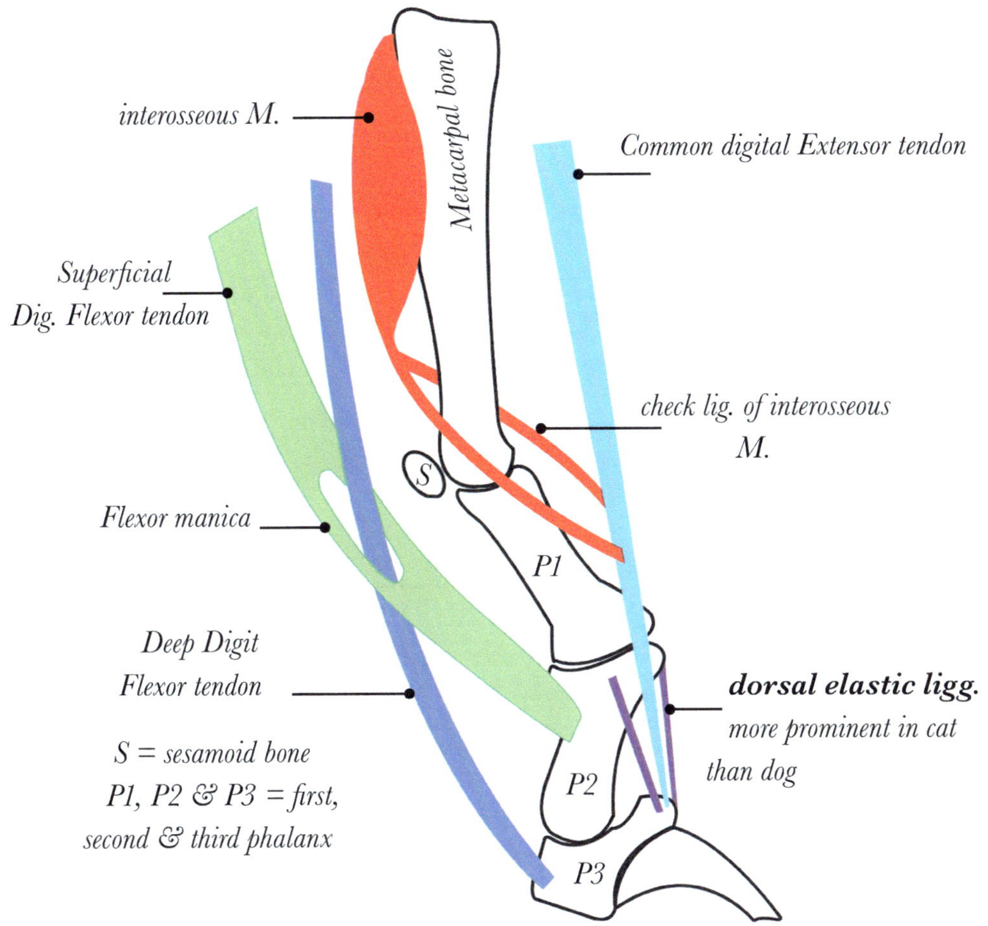

Cats can retract claws but dogs can not because:

Cats	Dogs
• Oblique articular surfaces of P2 & P3 allow rolling of P3 laterally alongside P2. • Strong dorsal elastic ligg. passively retract the claws. • DDF contraction extends (protracts) the claws.	• P3 cannot swing laterally • the dorsal elastic ligaments are are weak • the claws remain exposed.

Rule: Muscles on the flexor side of a joint flex that joint, while muscles on the opposite side extend it.

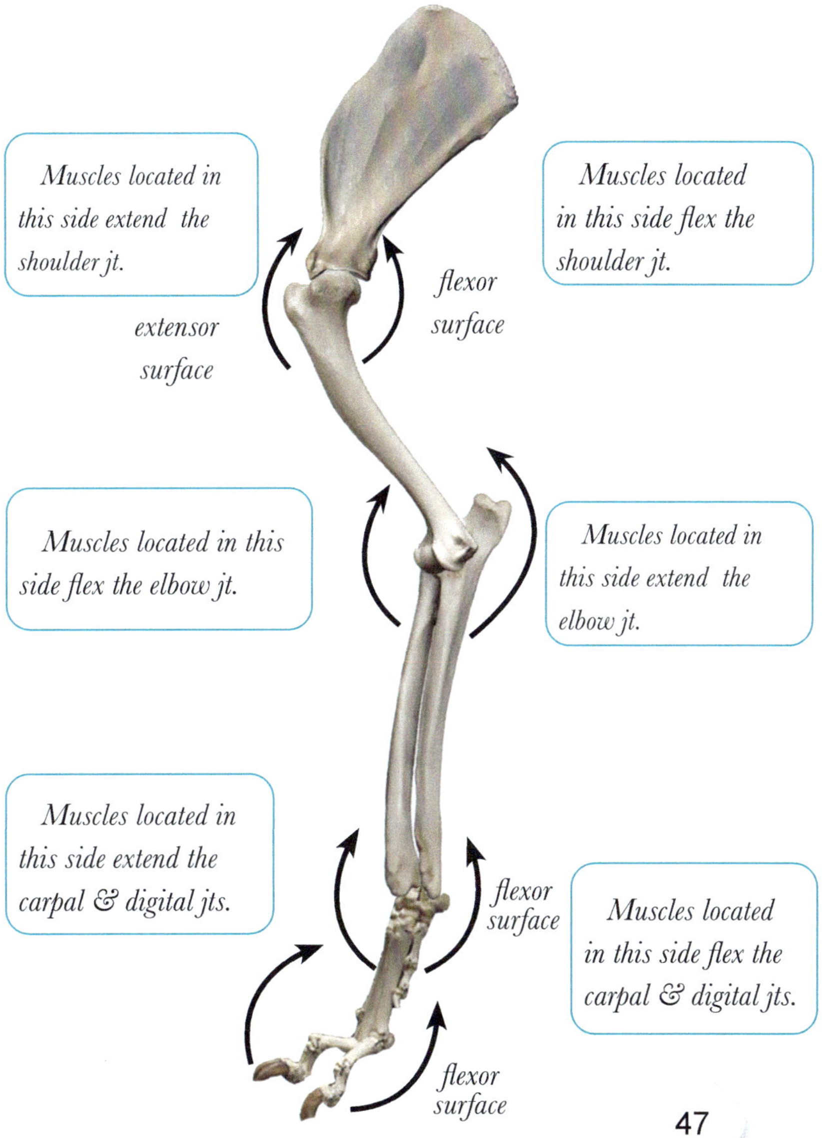

CHAPTER 5

CHAPTER 5

Arthrology of the Forelimb

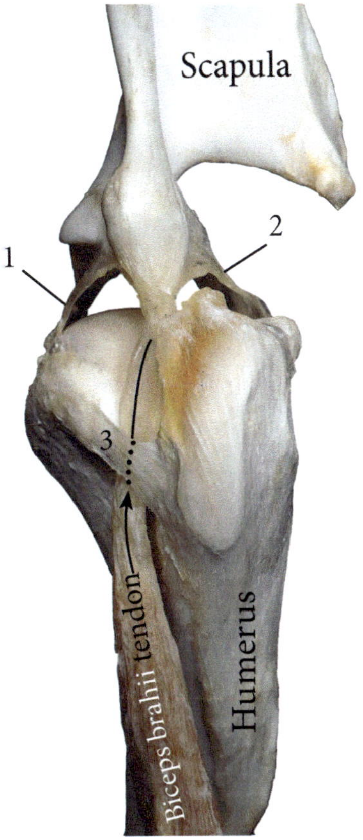

Lt. shoulder jt. - Cr. view

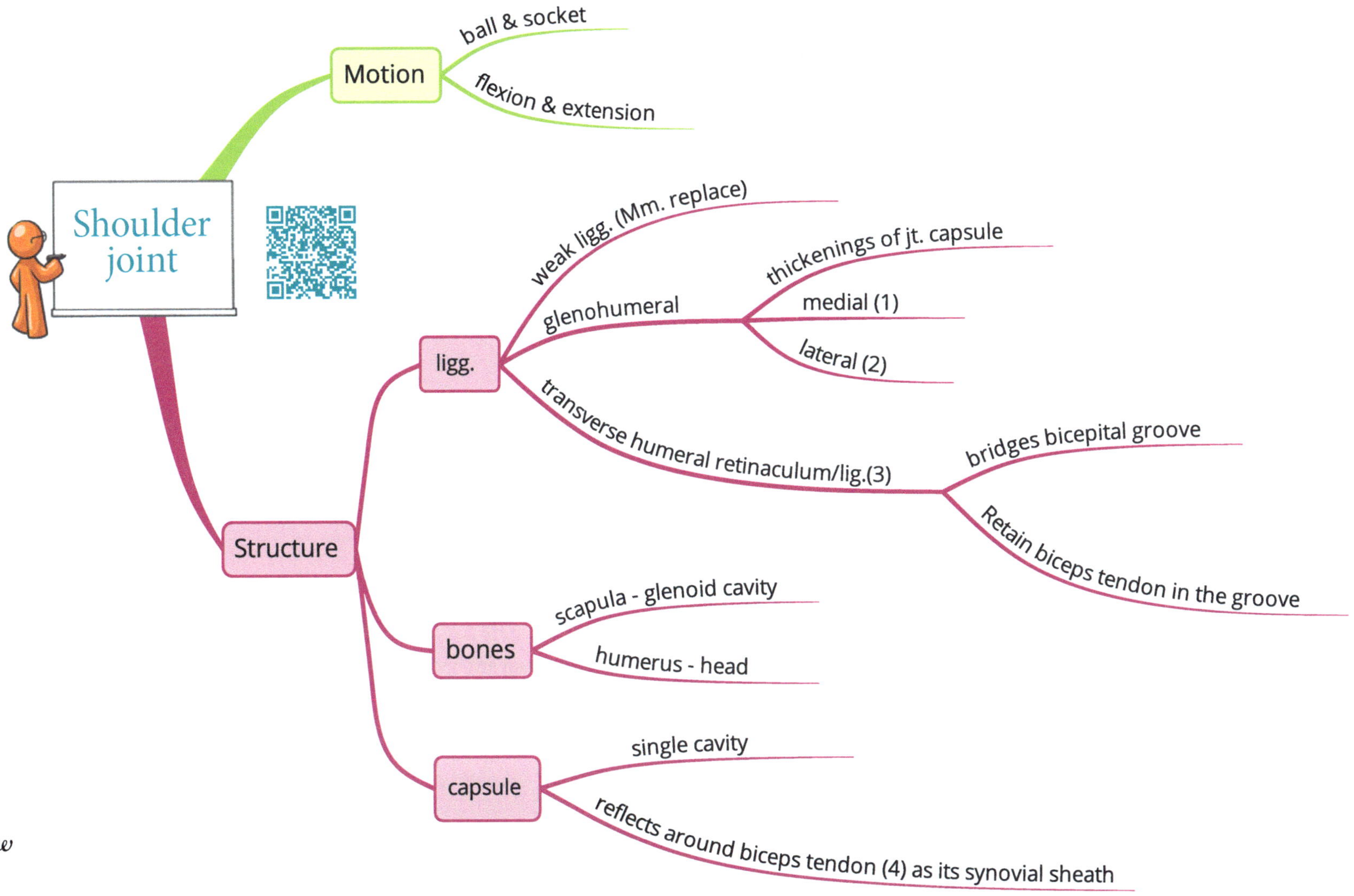

Why is shoulder jt. luxation & subluxation rare? Because it is protected by thick surrounding muscles & has a limited range of motion.

The acromion process & the greater tubercle of the humerus are landmarks for the shoulder jt.

Why inflammation & swelling of the biceps tendon leads to pain & lameness? Because the tendon is tightly enclosed by the transverse retinaculum within the intertubercular groove.

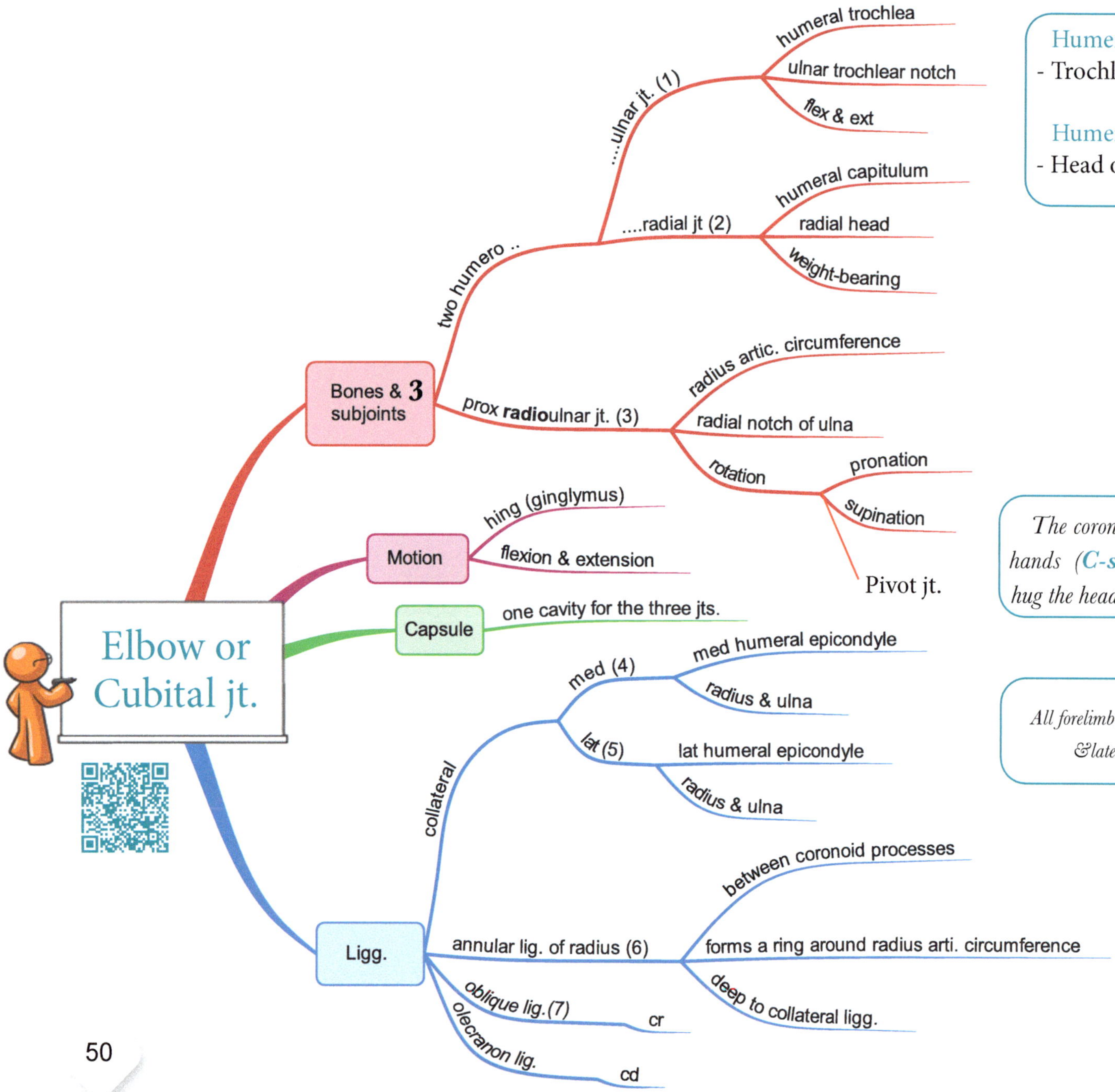

Humeroulnar: Trochlea Trochlea
- Trochlea of humerus + Tochlar notch of ulna

Humeroradial : Head Head
- Head of humerus (capitulum) + Head of radius

*The coronoid processes of the ulna send hands (**C-shaped annular lig.**) to hug the head of the radius.*

All forelimb jts. have collateral ligg. (medial &lateral) except the shoulder jt.

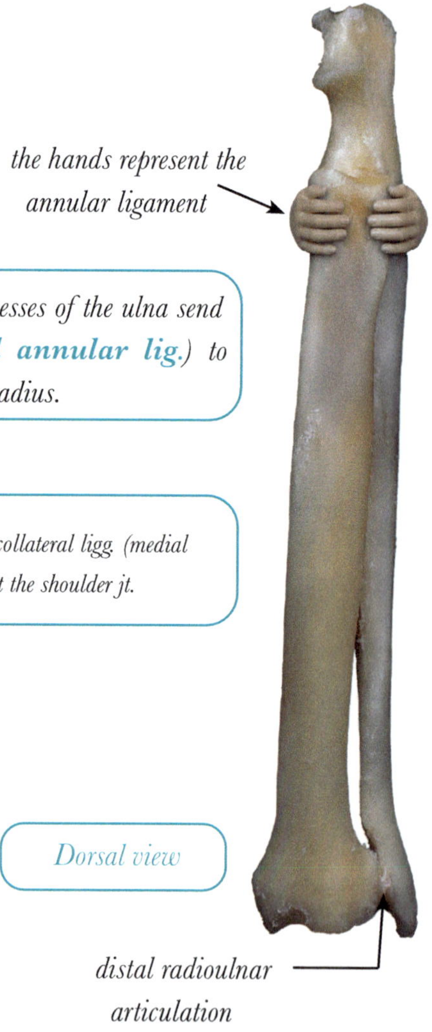

Elbow Joint

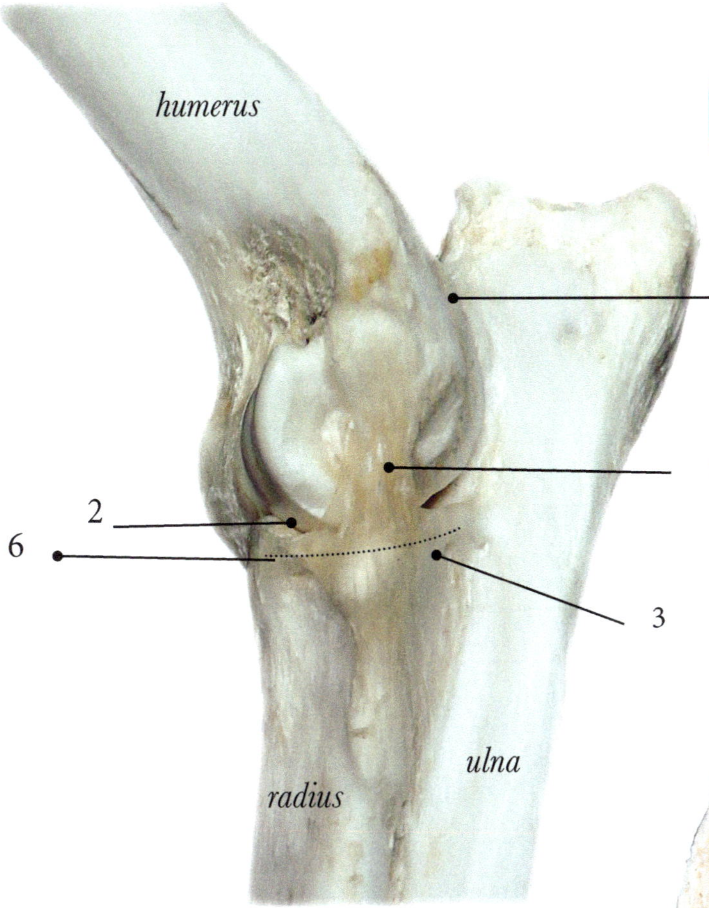

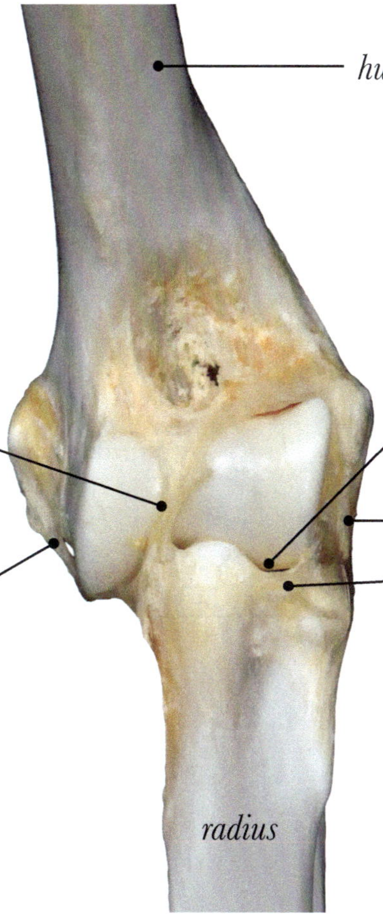

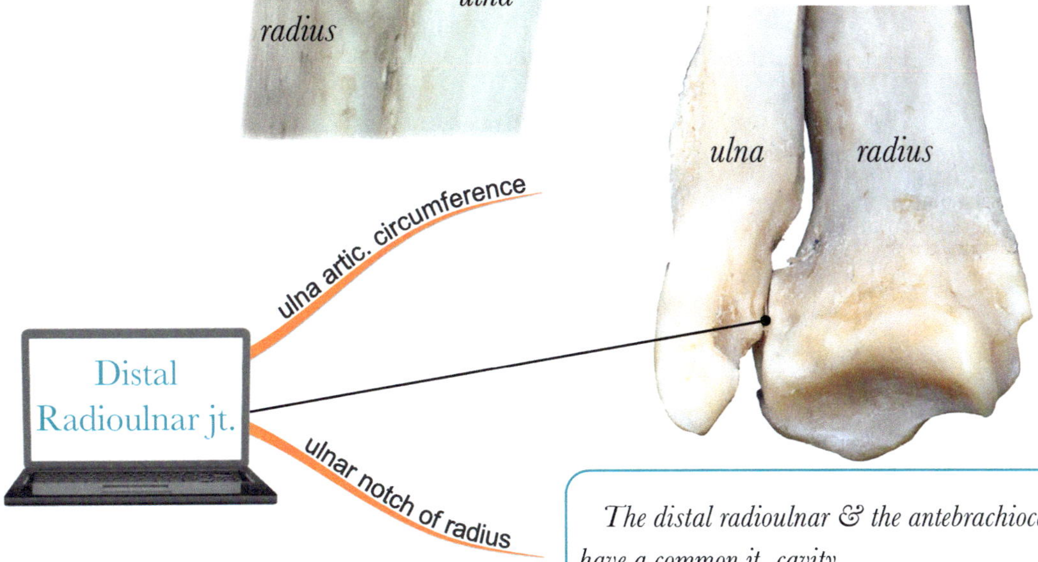

The distal radioulnar & the antebrachiocarpal jts have a common jt. cavity.

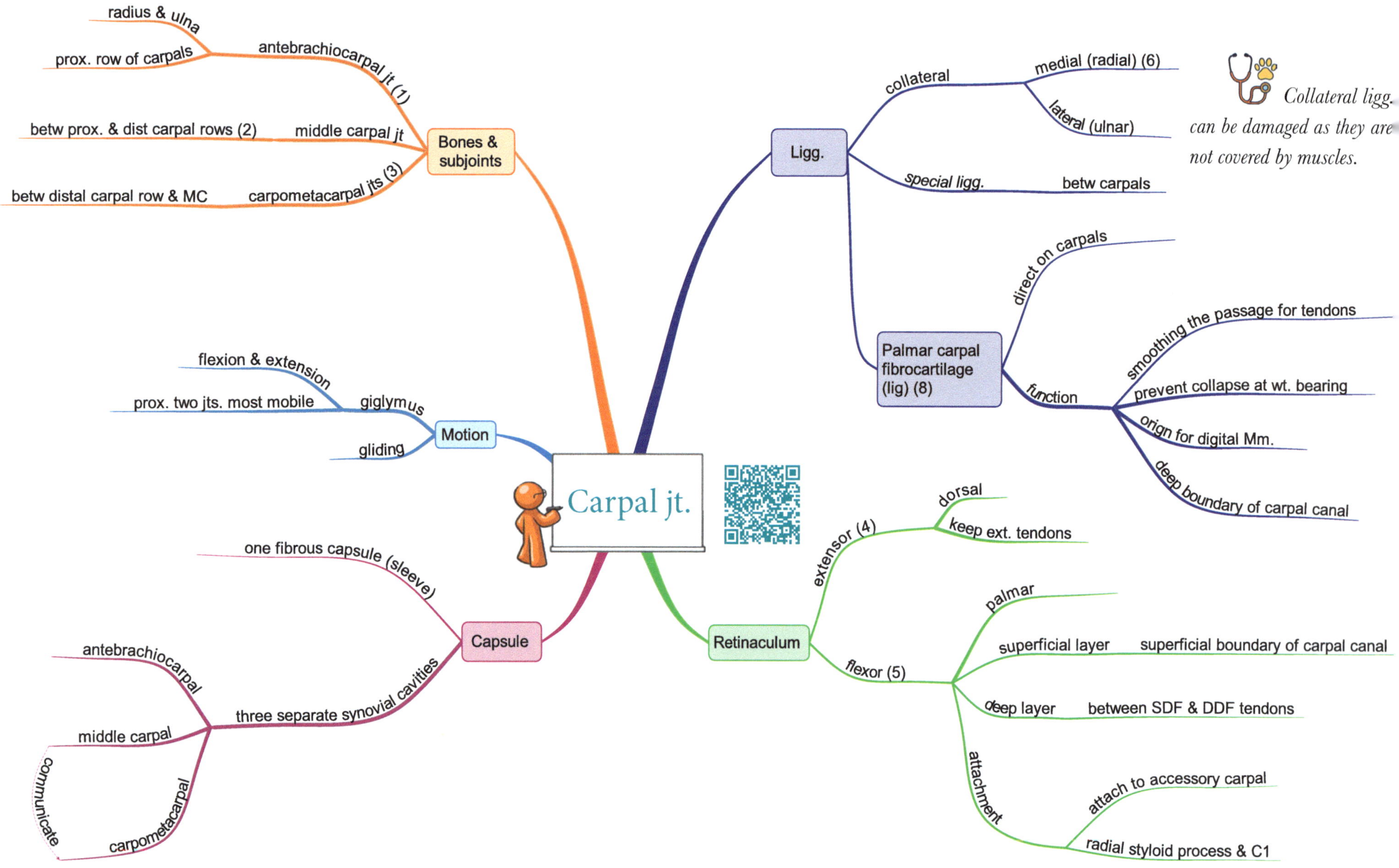

Collateral ligg. can be damaged as they are not covered by muscles.

Clinical note

Damage of palmar carpal lig. --> carpal hyperextension injury as when a dog jump from height or a car.

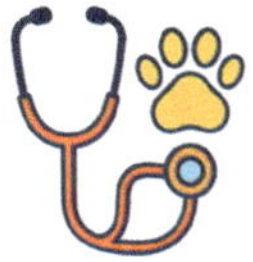

- *When the carpus is flexed, 90° angle,* ***the antebrachiocarpal & midcarpal joints*** *open & become easy to access for injection or fluid collection.*
- ***The carpometacarpal joint*** *is difficult to access directly; however, because it communicates with the middle carpal joint, separate injection is unnecessary. Injection of the middle carpal joint also reaches the carpometacarpal joint.*

Dorsal view

Palmar view

Metacarpal jts.

- **Intermetacarpal Jts.**
  - limited motion
  - betw MC bones (proximally)
- **Metacarpophalangeal Joints** *
  - **Bones**
    - head of MC
    - base of P1
    - three sesamoids per each jt.
      - one dorsal
      - two palmar (prox sesamoids)
  - **Motion**
    - ginglymus
    - flex - ext
  - **Capsule**
    - each jt. has its own capsule
  - **Ligg.**
    - collateral
      - medial
      - lateral
    - sesamoidean
      - dorsal
      - palmar

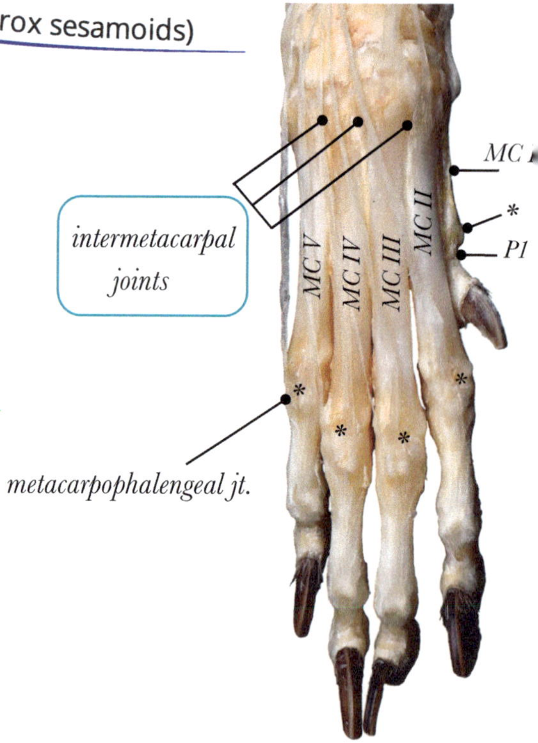

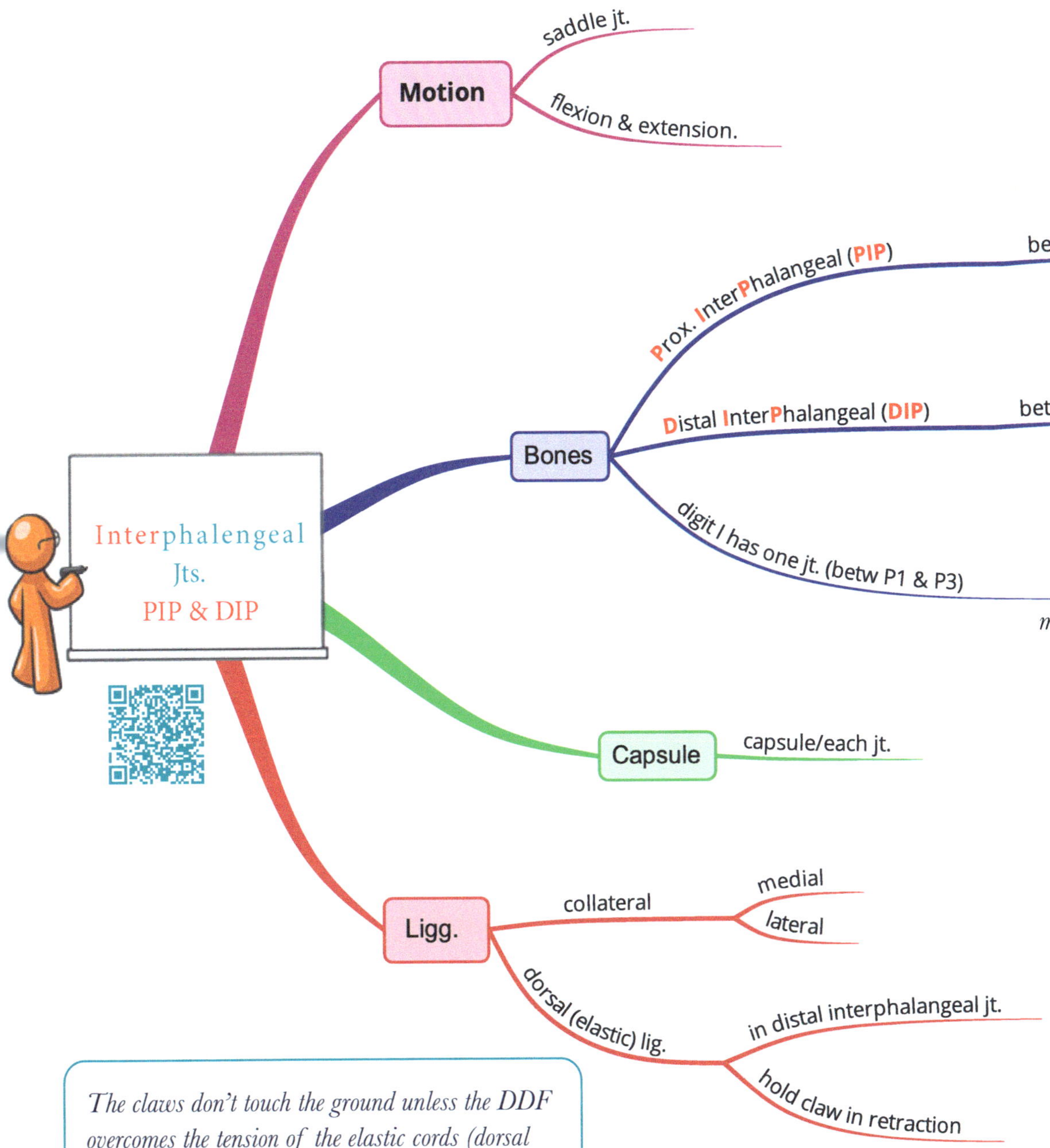

The claws don't touch the ground unless the DDF overcomes the tension of the elastic cords (dorsal ligg.) or the claws become overgrown.

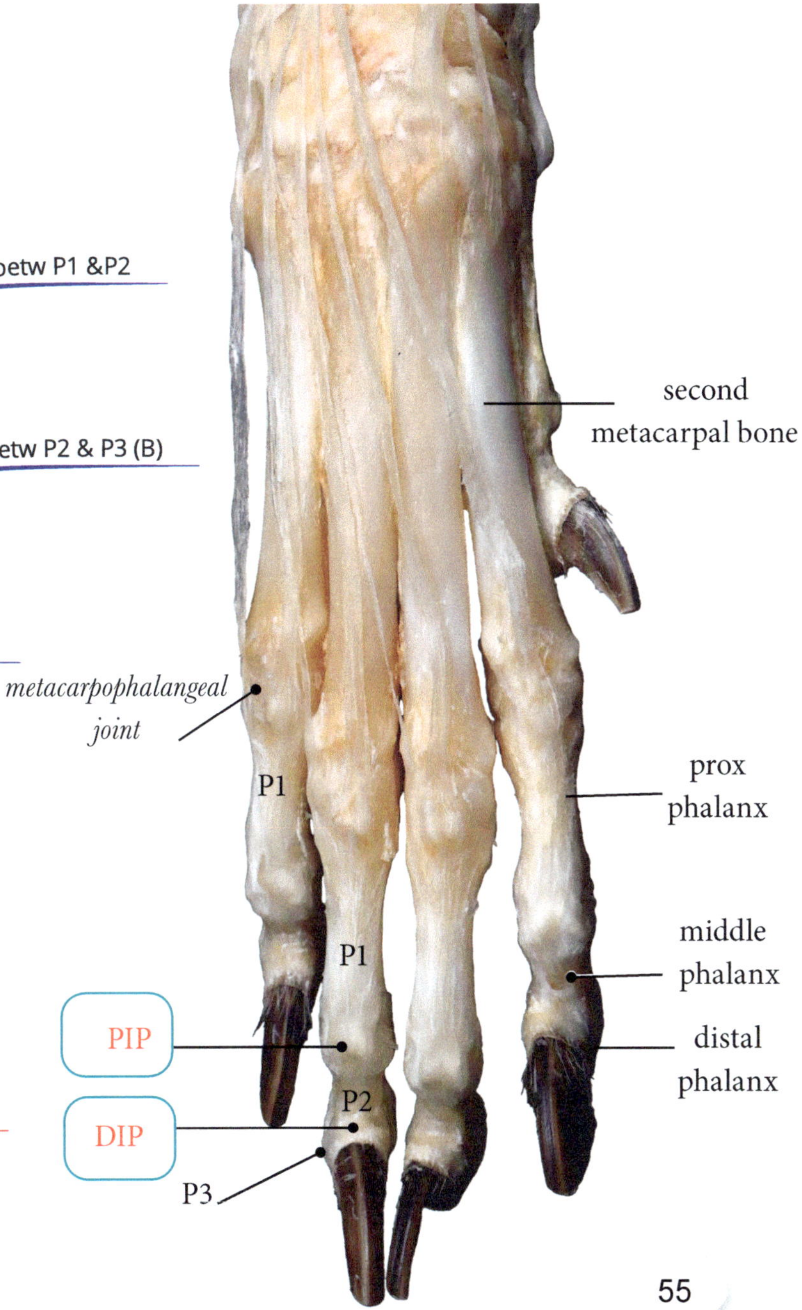

Rt. manus - Dorsal view

CHAPTER 6

CHAPTER 6

Myology of the Forelimb

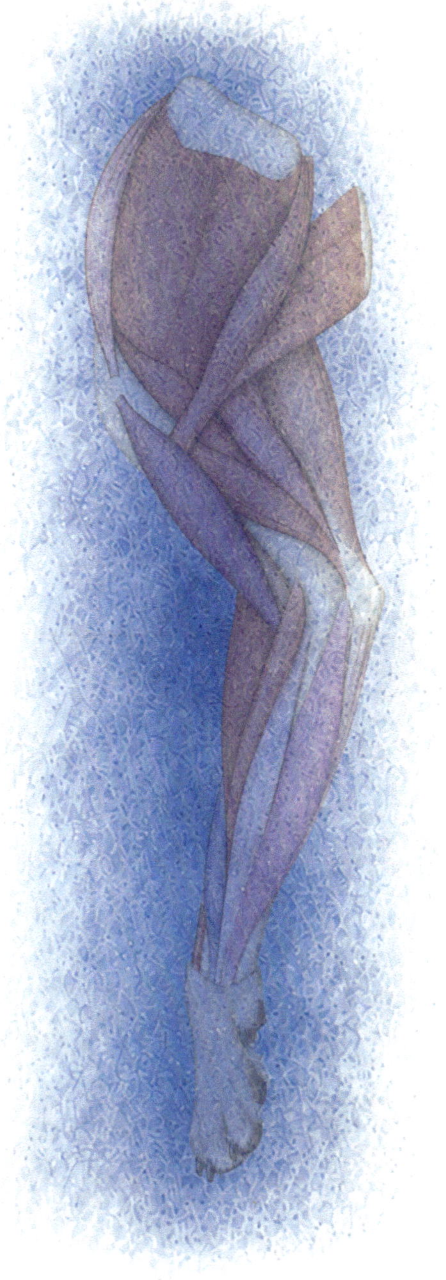

Synsarcosis & Extrinsic Mm. of forelimb

- Dorsal group
  - Latissimus dorsi
  - Trapezius
    - cervicis
    - thoracis
  - Rhomboideus
    - cervicis
    - thoracis
- Ventral group
  - Pectoral Mm.
    - superficial
      - cranial (descending)
      - caudal (transverse)
    - deep
      - cranial (subclavius) — *absent in carnivores*
      - caudal (ascending)
  - Brachiocephalicus
  - Serratus ventralis
    - cervicis
    - thoracis
  - Omotransversarius

Cats' cleidobrachialis attaches to the ulna with the superficial pectoral (more adduction --> facilitate climbing). In dog, attached to cr. border of humerus.

Extrinsic muscles of the forelimb

Dorsal group

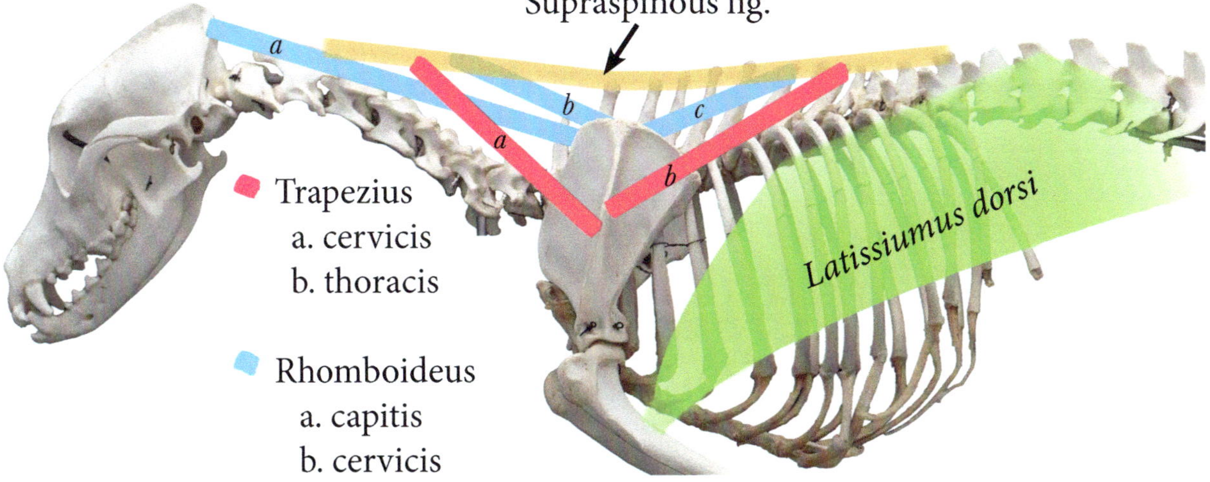

Knowledge of the extrinsic forelimb muscles is essential for forelimb amputation.

Ventral group

Omo = *shoulder*, *Transversarus* = *transverse*
→ *so it extends from the shoulder (acromion) to the transverse process (wing) of the atlas.*

Omotransvarsarius muscle
Covers superficial cervical Lnn.

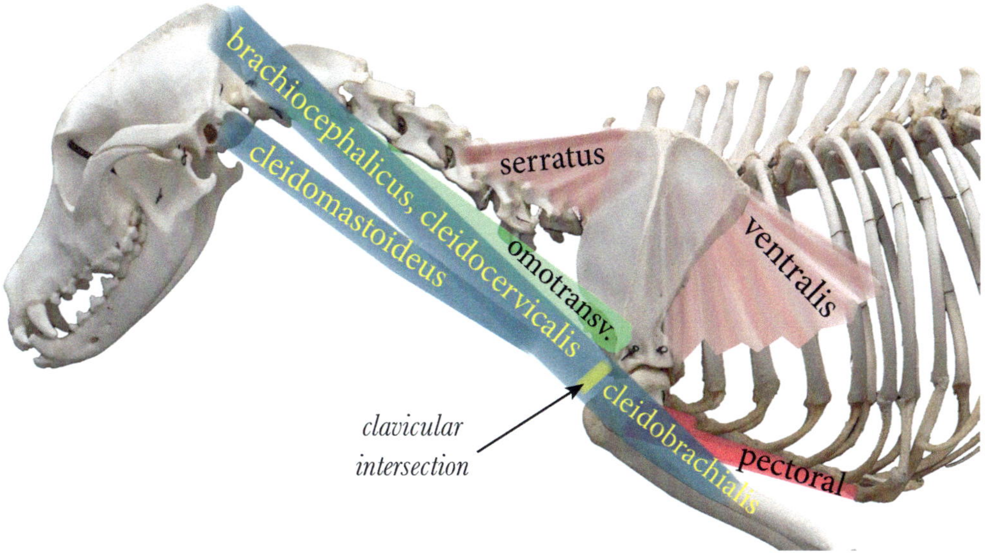

Cleido = *clavicle*
Brachialis = *arm*
→ *so it extends from the clavicular intersection to the arm*

Forelimb Intrinsic Mm. (Shoulder)

Surgical approach to the humeral head: *between the acromial part of the deltoideus and the supraspinatus.*

the subscapularis M. acts as a medial collateral ligament of shoulder jt.

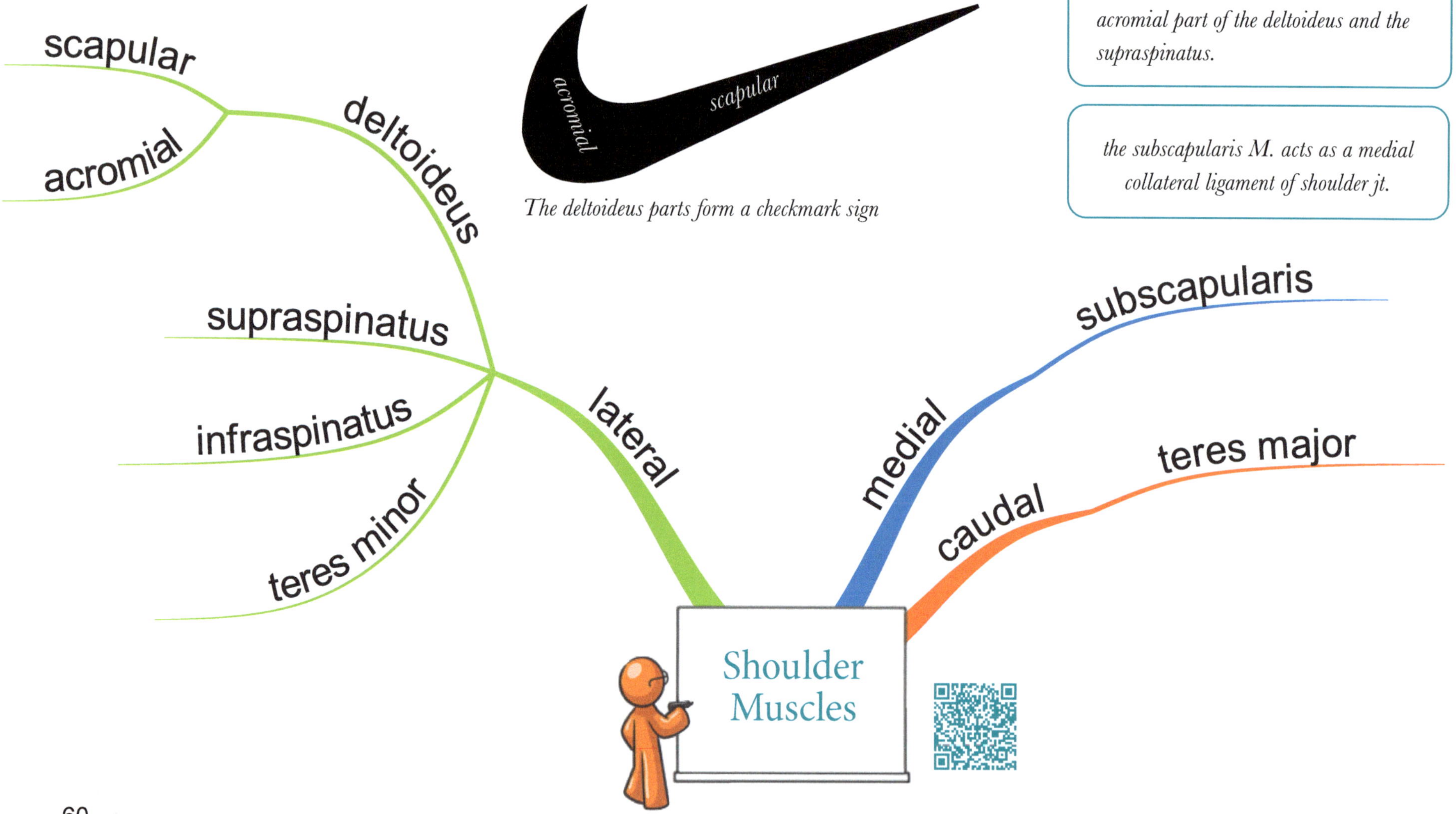

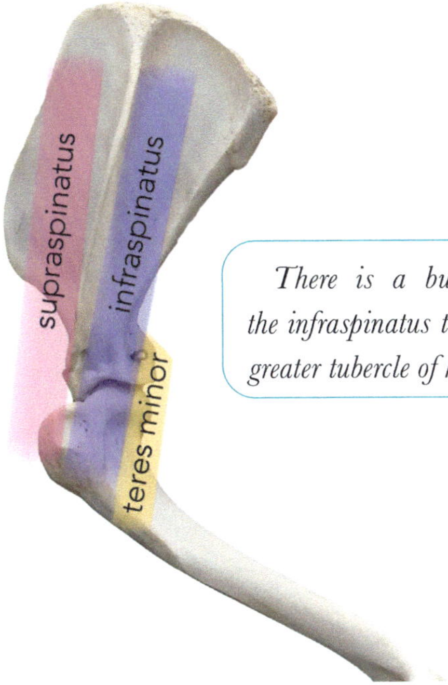

There is a bursa between the infraspinatus tendon & the greater tubercle of humerus

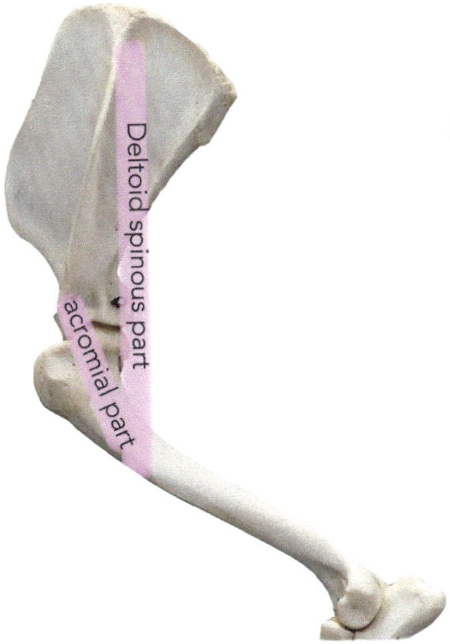

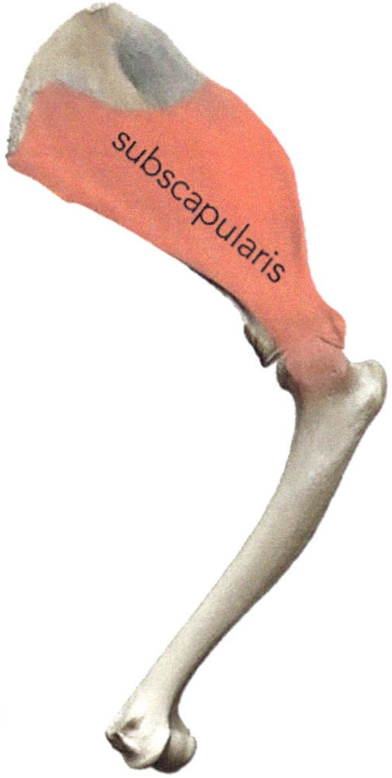

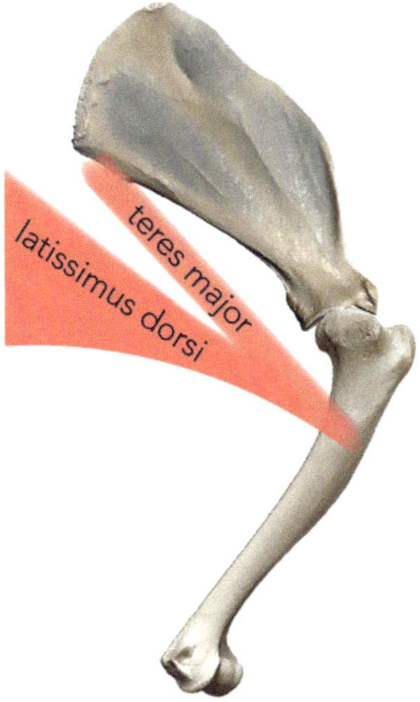

Muscle	***Origin***	***Insertion***	***Action***	***Innervation***
Supraspinatus	*supraspinous fossa*	*greater tubercle of humerus*	*stabilize & extend shoulder*	*suprascapular n.*
Infraspinatus	*infraspinous fossa*		*stabilize, extend & flex shoulder*	
Teres minor	*infraglenoid tuber & cd. border of scapula*	*teres minor tuber*	*shoulder flexion*	
Deltoideus: ***scapular part*** ***acromial part***	*- scapular spine* *- acromion*	*deltoid tuberosity*	*shoulder flexion*	*axillary n.* *axillary n.*

Muscles of Shoulder & Arm, Lateral view

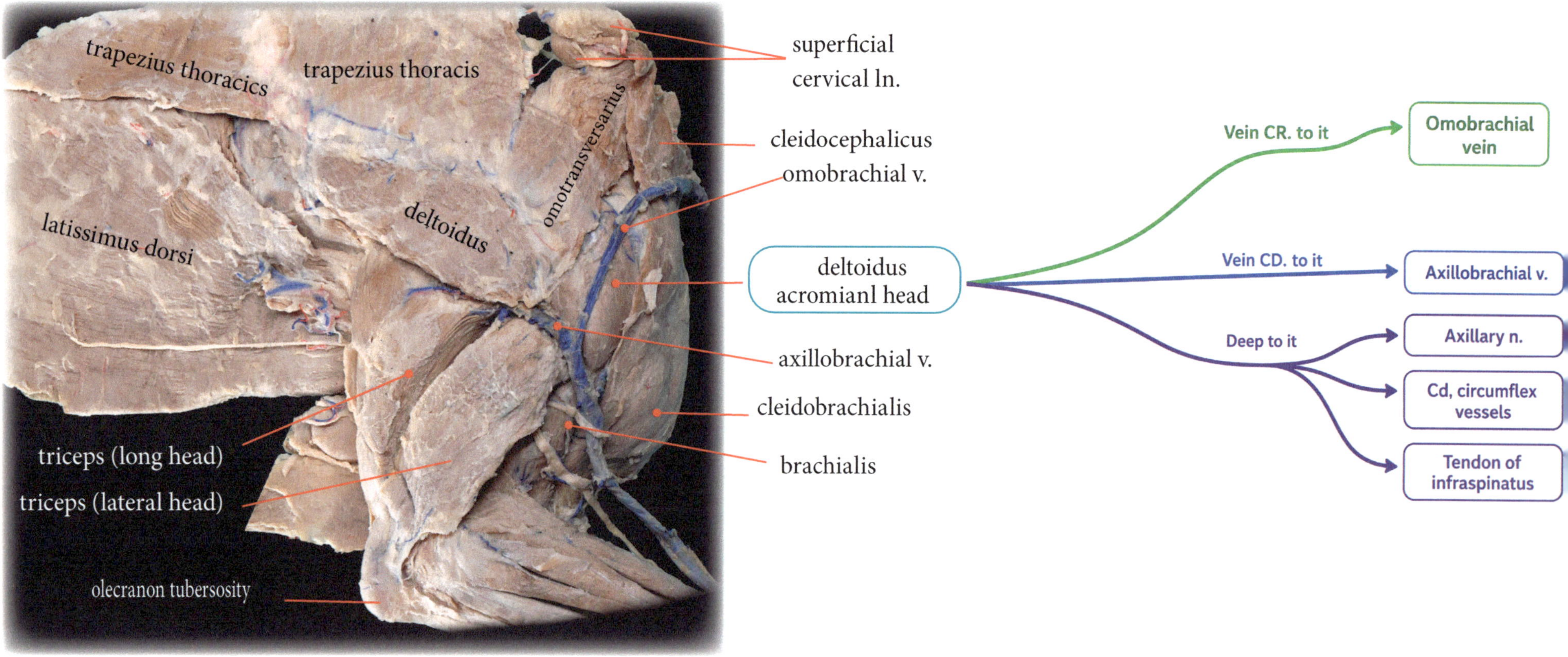

Muscle	Origin	Insertion	Action	Innervation
Subscapularis	*subscapular fossa*	*lesser tubercle of humerus*	*stabilize, adduct & extend shoulder*	***SUBSCAPULAR N.***+ *axillary n.*
Teres major	*caudal scapula*	*teres major tuberosity*	*shoulder flexion*	*axillary n.*
Latissimus dorsi	*thoracolumbar fascia*			*thoracodorsal n.*

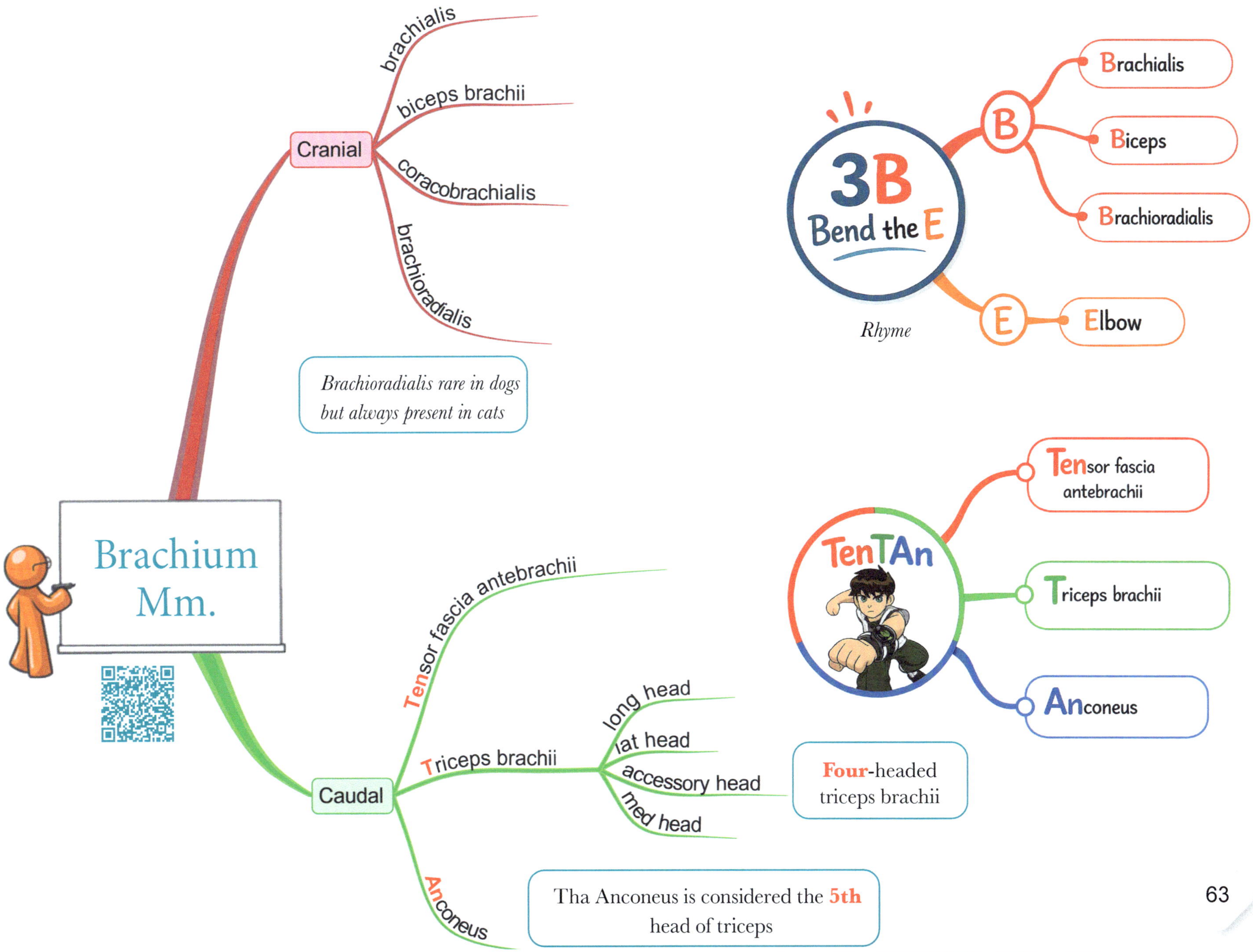
Brachium Mm.
Cranial
brachialis
biceps brachii
coracobrachialis
brachioradialis
Brachioradialis rare in dogs but always present in cats
Caudal
Tensor fascia antebrachii
Triceps brachii
long head
lat head
accessory head
med head
Four-headed triceps brachii
Anconeus
Tha Anconeus is considered the 5th head of triceps
3B Bend the E
Rhyme
B
Brachialis
Biceps
Brachioradialis
E
Elbow
TenTAn
Tensor fascia antebrachii
Triceps brachii
Anconeus

Triceps Brachii

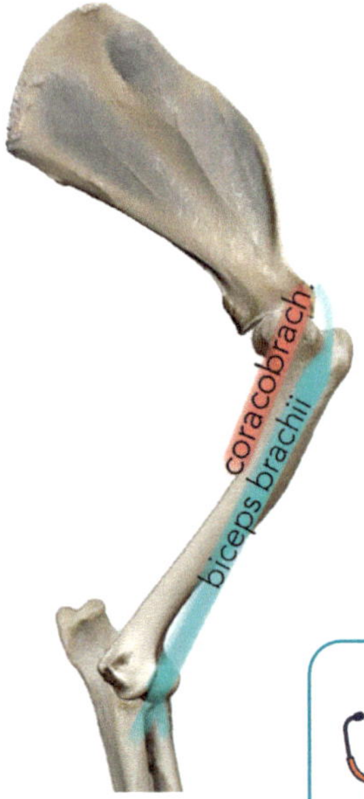

The brachialis M. "breaks" between the bi & triceps brachii.
The brachialis tendon "breaks" between the 2 the insertions of the biceps.

Rupture of Biceps brachii tendon (or avulsion of the supraglenoid tuber.) → shoulder over flexion + elbow overextension.

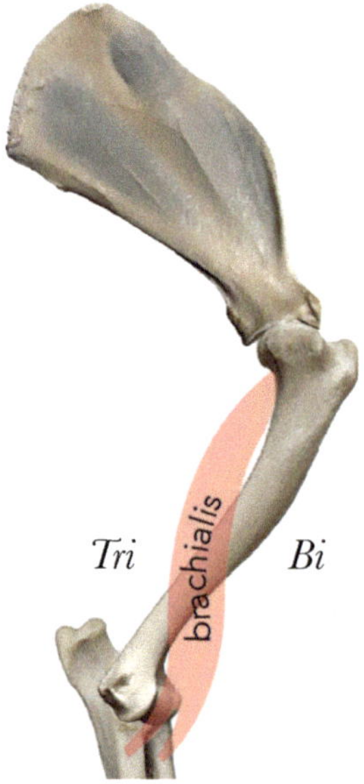

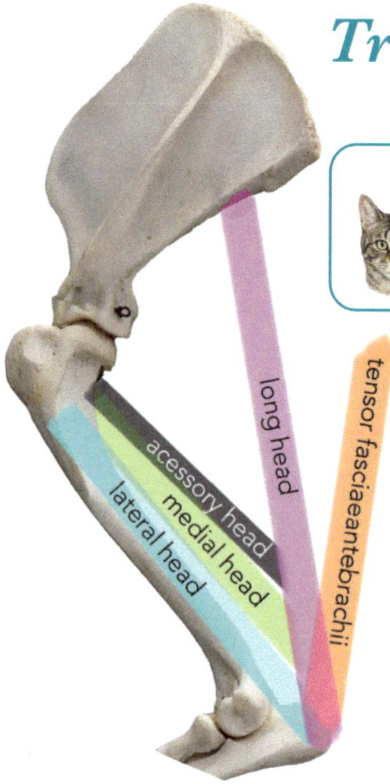

Medial & Accessory heads of Triceps are fused in Cats.

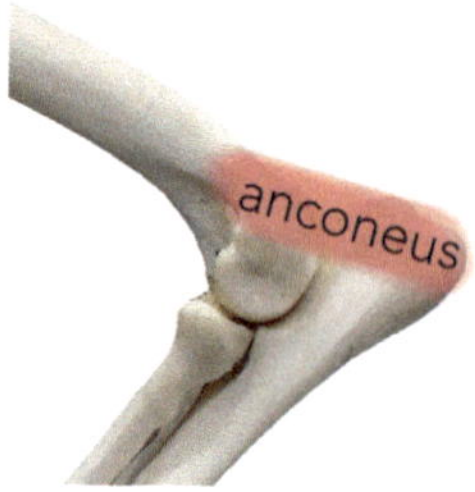

The triceps M. fills the triangle between scapula, humerus & olecranon. Its caudal contour forms the ***tricipital margin****; a key auscultation landmark.*

- *Rupture of the tendon of the triceps prevents elbow extension .*
- ***Subtendinous bursa of triceps*** *between tendon of triceps & olecranon.*
- *Elevate anconeus M. to remove the ununited anconeal process* ***(UAP)****.*

<table>
<tr><th>Brachium Mm.</th><th>Origin</th><th>Insertion</th><th colspan="2">Action</th><th>Innervation</th></tr>
<tr><td>Brachialis</td><td>prox. lat. humerus</td><td>Ulnar tuberosity</td><td rowspan="2">flexes Elbow</td><td></td><td rowspan="3">BBC Nerve
B B C</td></tr>
<tr><td>Biceps brachii</td><td>supraglenoid tuber</td><td>Radial & ulnar tuberosities</td><td>extends shoulder</td></tr>
<tr><td>Coraco brachialis</td><td>coracoid process</td><td>Lesser tubercle of humerus</td><td colspan="2">adducts & extends shoulder jt.</td></tr>
<tr><td>Brachioradialis</td><td>lat. humeral epicondyle</td><td>Distal radius</td><td colspan="2">flex elbow
rotate radius laterally (supination)</td><td rowspan="4">Radial n.</td></tr>
<tr><td>Tensor fascia antebrachii</td><td>fascia covering latissimus dorsi</td><td rowspan="3">Olecranon tuber</td><td colspan="2">extends elbow & tenses antebrachial fascia</td></tr>
<tr><td>Triceps brachii
4 heads</td><td>cd. scapula (long head)
& proximal humerus</td><td colspan="2">flex shoulder (long head)
extends elbow</td></tr>
<tr><td>Anconeus</td><td>superacondylar crest,
medial & lat. epicondyles</td><td colspan="2">extends elbow</td></tr>
</table>

A Cat Cannot
Buy (Biceps)
from You (Ulna)

Meaning:
Cat biceps has no attachment to ulna.

Cat Brachialis Breaks
→
its attachment to Radius.

Meaning:
Cat brachialis has no attachment to Radius.

Note

In the dog:

*Both the **brachialis & biceps brachii** insert by two tendons; one on the ulnar tuberosity and one on the radial tuberosity.*

In the cat:

***Brachialis** is tendinous & has no radial attachment (inserts on ulna medial coronoid process of ulna only).*

***Biceps brachii** has no ulnar attachment.*

***Triceps medial & accessory** heads are fused.*

Muscles of Forearm

Craniolateral group

- **superficial**
  - *(4 muscles)*
    - 2 Carpal (at the prephary)
    - 2 Digital (in the center)
  - ext. carpi radialis
  - common digital extensor — to digits 2,3,4,5
  - lateral dig. extensor — to digits 3,4, 5
  - ulnaris lateralis — is named also ext. carpi ulnaris
- **deep**
  - *directly on bone*
  - supinator
    - *Supinator may contain an inconstant Sesamoid bone in its tendon of origin*
  - abductor pollicis longus — *contains a sesamoid @ its insertion @ the carpus*

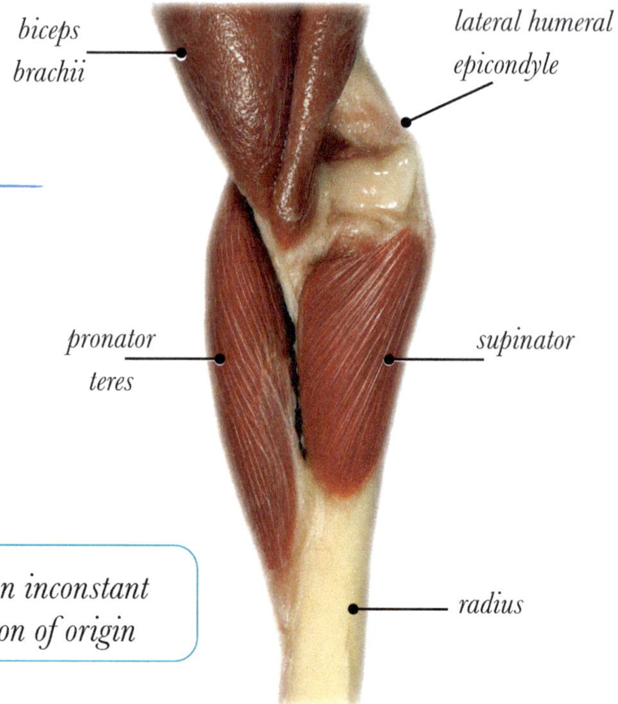

Muscles of forearm
Craniolateral Group

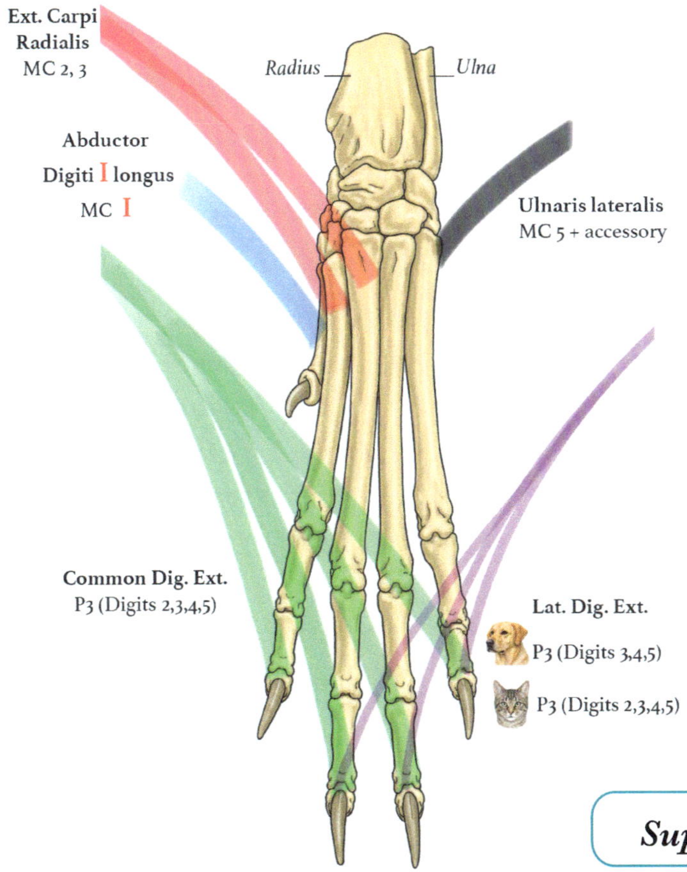

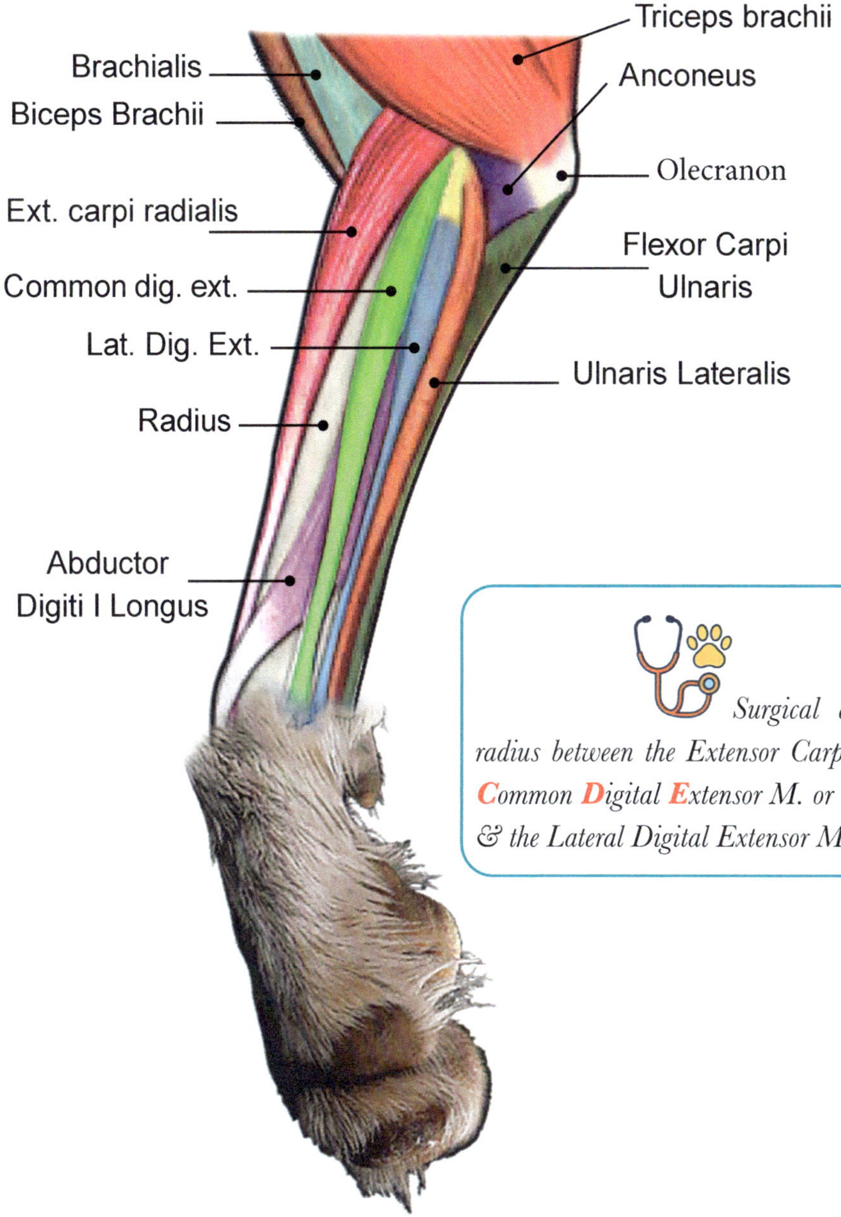

*Surgical approach to shaft of radius between the Extensor Carpi Radialis M. & the **C**ommon **D**igital **E**xtensor M. or between the **CDE** M. & the Lateral Digital Extensor M.*

Superficial muscles of the antebrachium, lateral view

Every Cow Loves You For Finding Some Fresh Pasture.
E
Extensor Carpi Radialis
C
Common digital extensor
L
lateral digital extensor
You
Ulnaris lateralis
For
Flexor Carpi ulnaris ulnar head
Finding
Flexor Carpi ulnaris humeral head
Some
Superficial digital flexor
Fresh
Flexor Carpi Radialis
Pasture
Pronator teres
YOU mark the change!
← Radial N. | Ulnar & Median Nn. →
FRESH PASTURE

Muscles of Forearm

Craniolateral group	**Origin**	**Insertion**	**Action**	**N.**
Ext. Carpi Radialis	*Epicondylar crest + Lat. Humeral Epicondyle*	*Prox. ends of* ***MC*** *2 & 3*	*Extends the carpus, flexes the elbow*	**Radial n.**
Common Dig. Evxtensor	**Lat. Humeral Epicondyle**	*P3 extensor processes of* ***Digits*** *2-5*	*Extends carpus & digit*	
Lateral Dig. Extensor		*P3 extensor processes of* ***Digits*** *3-5 (the* ***lateral*** *digits)*		
Ulnaris Lateralis		*Accessory carpal bone & prox* ***MC*** *5*	*flex the carpus*	
Supinator		*Cr. proximal fourth of radius*	*Rotation (****Supination****) of antebrachium laterally so the paw palmar surface turns up.*	
Extensor Carpi Obliques aka: Abductor digiti I (Pollicis) longus	*Craniolateral shaft of radius & ulna*	*Prox. end of* ***MC1***	*Aabducts digit 1* *Extends the carpus*	

Notice:

Most of this craniolateral antebrachial group originate from the ***t*** *humeral epicondyle and insert in MC or Digits.*
The ulnaris lateralis M. is aka. Extensor carpi ulnaris. It is the the most lateral of the group. It is morphologically extensor but physiological flexor.

Muscles of Forearm

Caudomedial group

- superficial
  - pronator teres (1)
  - flexor carpi radialis (2)
  - superficial digital flexor (3)
  - flexor carpi ulnaris
    - ulnar head (4)
    - humeral head
- deep
  - deep dig. flexor
    - humreal head
    - radial head
    - ulnar
- deepest
  - pronator quadratus

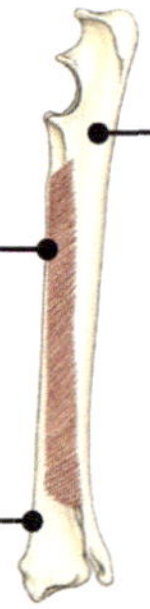

FM radio i.e. the Flexors are Medial & the origin of the Flexors is from the Medial humeral epicondyle while the extensors are lateral & originate from the lateral humeral epicondyle.

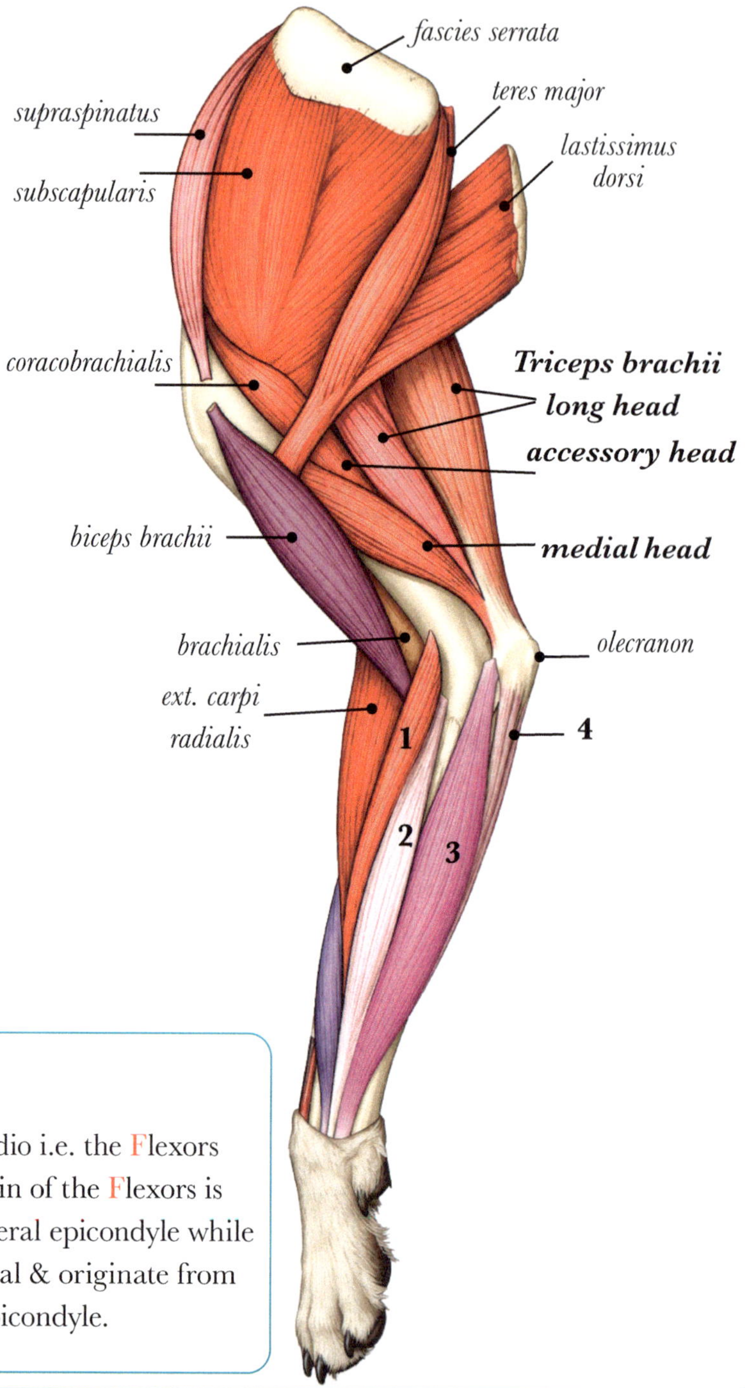

Muscles of Forearm

Caudomedial group	***Origin***	***Insertion***	***Action***	***N.***
Flexor Carpi Radialis	*medial humeral epicondyle*	*palmar bases of* ***MC*** *2, 3*	*flexes the carpus*	*Median n.*
Flexor Carpi Ulnaris	***Humeral head:*** *medial humeral epicondyle* ***Ulnar head:*** *Olecranon*	*accessory carpal bone*	*flexes the carpus*	*Ulnar n.*
Superficial Digital Flexor	*medial humeral epicondyle*	*middle phalnages of digits 2-5*	*flexes:* *carpus* *metacarpophalangeal jts.* *prox. interphalangeal jts.*	*Median n.*
Deep Digital Flexor	***Humeral head*** *.... medial humeral epicondyle* ***Ulnar head*** *.... cd. ulnar border* ***Radial head*** *.... prox. medial radius*	*flexor tubercle of P3 of digits 1-5*	*flexes carpus & digital*	*Median & Ulnar Nn.*
Pronator teres	*medial humeral epicondyle*	*prox. radius*	*Rotation (**Pronation**) of antebrachium medially so the paw turns palmarly (down)* *Flexes the elbow*	*Median n.*

Deep Digital Flexor Muscle

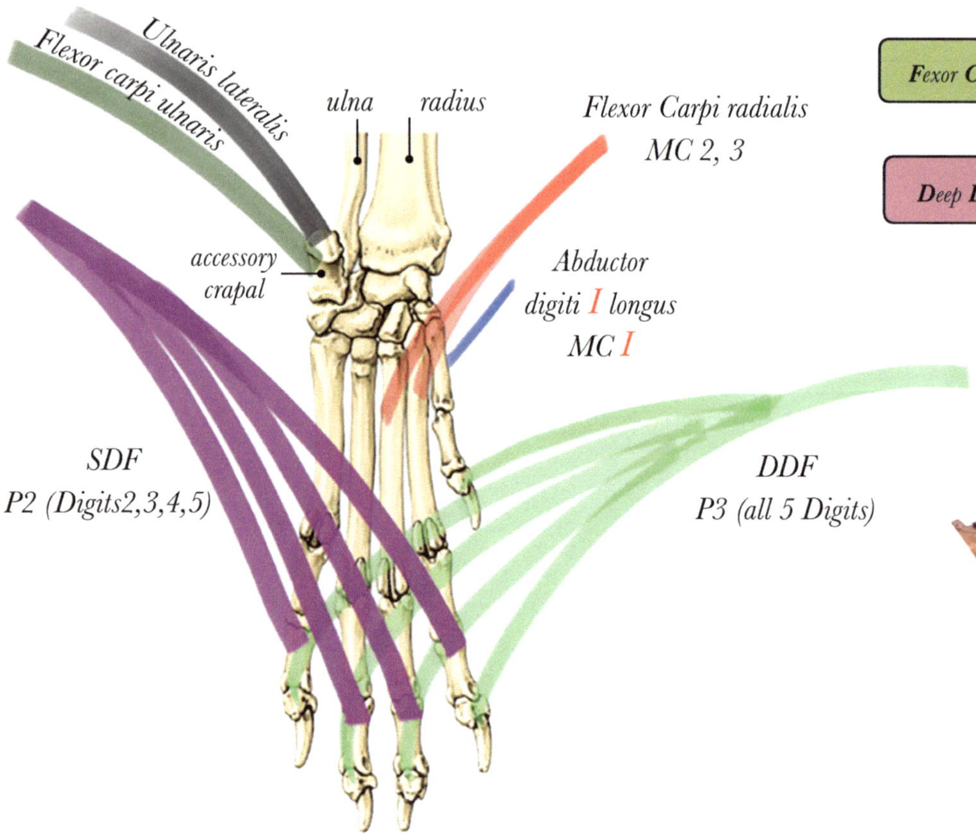

Insertion of forearm muscles
(left manus, palmar view)

Left antebrachial Mm dissected (medial view)

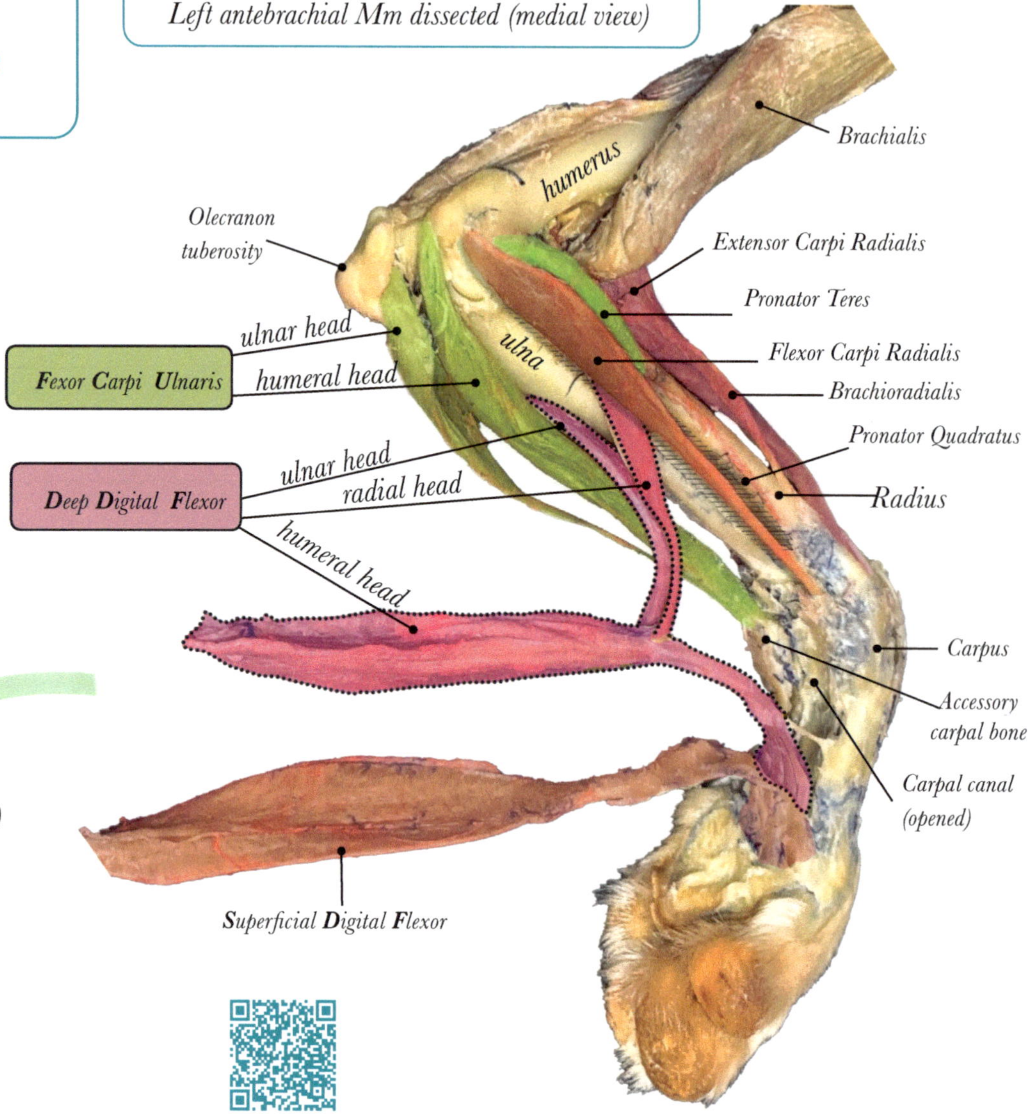

The tendons of the 3 heads unite to form the DDF tendon, which passes through the manica flexoria of the SDF tendon.

Carpal Canal

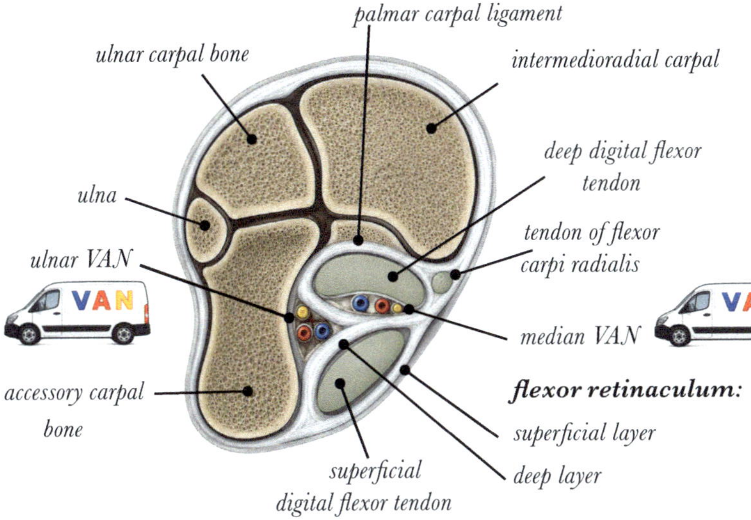

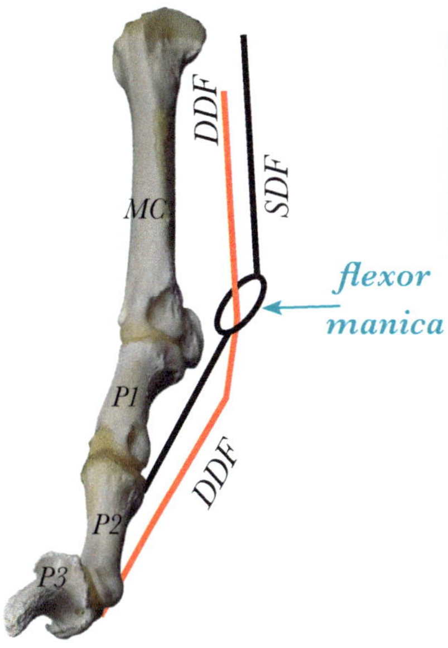

Flexor manica (manica flexoria):

Since the SDF ends at P2 and the DDF needs to reach its distenation (P3). So each DDF tendon "perforates" the SDF to reach to P3 (this happen at the level of the MCP jt.) So SDF becomes like a tube enclosing the DDF (this tube is called ***flexor manica****). distally the formed tube appears palmarly split into two branches with the DDF in between. The 2 SDF splits end in P2.*

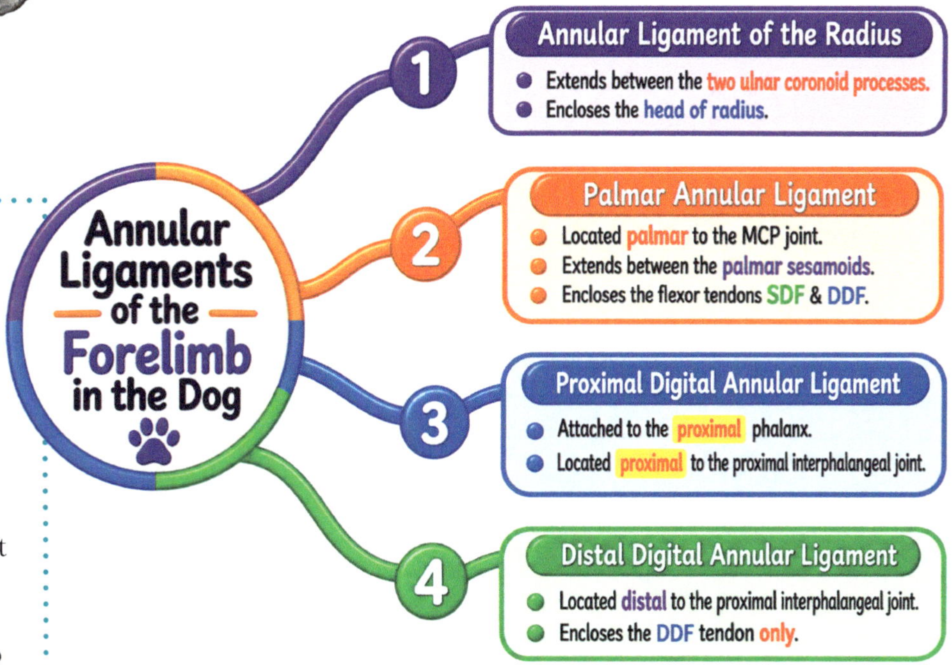

- Carpal canal = carpal groove + flexor retinaculum
- SDF & DDF tendons pass in the carpal canal
- flexor retinaculum has superficial & deep layers
- SDF & DDF tendons for each digit are bounded down by palmar annular lig. & prox. & distal dig. annular ligg.
- All digital tendons (extensors & flexors) split into branches that insert in the digits.
- Each SDF tendon ends at P2, so it splits (forming manica flexoria) to let the DDF tendon reaches to P3.
- ***SDF splits into two for DDF pass through. (rhyme).***

*"**The extensor that flexes.**" is the ulnaris lateralis (extensor carpi ulnaris), this is a misleading name, as it flexes the carpus.*

Forelimb Muscle Actions

- **Shoulder jt.**
  - cr (extensors)
    - supraspinatus
    - coracobrachialis
  - cd (flexors)
    - infraspinatus
    - deloideus
    - teres major & minor
- **Elbow jt.**
  - cr (flexors)
    - biceps brachii
    - brachialis
    - supinator
    - pronator teres
  - cd (extensors)
    - tensor fasciae antebrachii
    - triceps brachii (all heads)
    - anconeus
- **Carpus, MC & digits**
  - cr (extensor)
    - ext. carpi radialis
    - Abductor digiti I longus
    - digital extensor
      - common
      - lateral
  - cd (flexors)
    - ulnaris lateralis
    - flexor carpi radialis
    - superficial digital flexor
    - deep digital flexor
- **Pronators of forearm**
  - pronator teres
  - pronator quadratus

Separate the pronator teres from the flexor carpi radialis (medial approach to elbow jt.) to remove fragmented medial coronoid process.

Separate the ext. carpi radialis from the common dig ext. to expose the radius for surgery.

Separatie the ulnaris lateralis from the flexor carpi ulnaris to expose the ulna for surgery,

Forelimb Muscle Actions

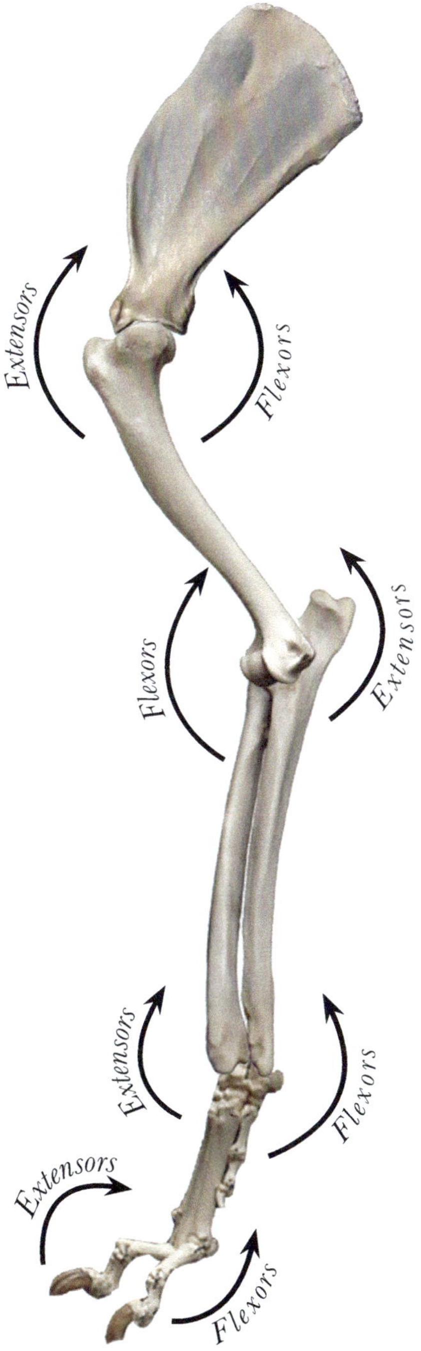

Extensors

Subscapularis
Supraspinatus
Infraspinatus
Biceps brachii
Coracobrachialis

Flexors

Infraspinatus
Triceps brachii (long head)
Teres major
Deltoideus

Flexors

Biceps Brachii
Brachialis
Brachioradialis (constant in cats)
Pronator teres
Supinator
Extensor carpi radialis

Extensors

Triceps brachii (all heads)
Tensor fasciae antebrachii
Anconeus

Extensors

Extensor carpi radialis
Common digital extensor
Lateral digital extensor
Abductor pollicis longus

Flexors

Flexor carpi radialis
Flexor carpi ulnaris
SDF DDF
Ulnaris lateralis

Extensors

Common dig. extensor
Lateral digital extensor

Flexors

SDF (digits 2-5)
DDF (digits 1-5)

CHAPTER 7

Neurovascular Anatomy of the Forelimb

REGIONAL BLOOD VESSELS, LYMPHATIC STRUCTURES, AND THE BRACHIAL PLEXUS

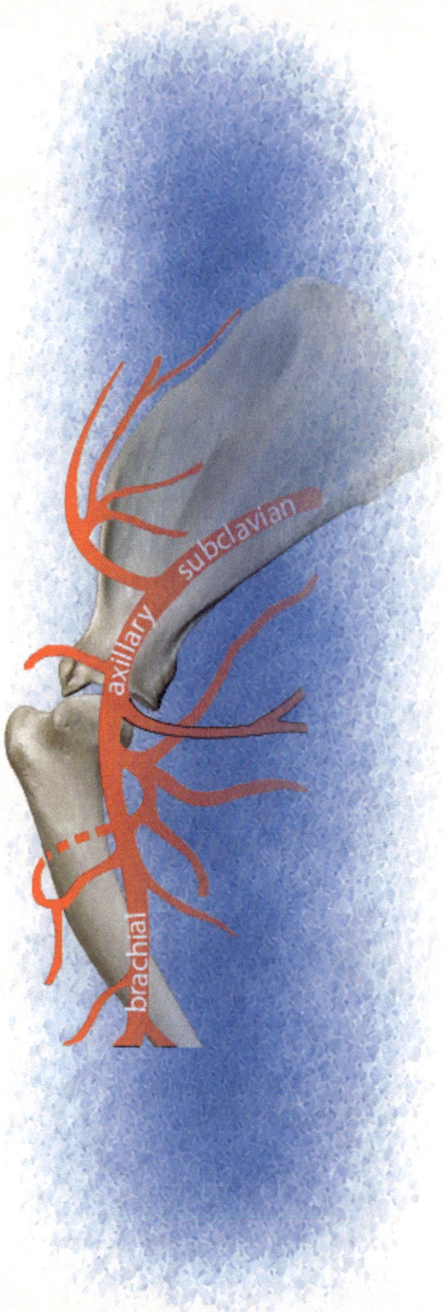

Arteries of Forelimb

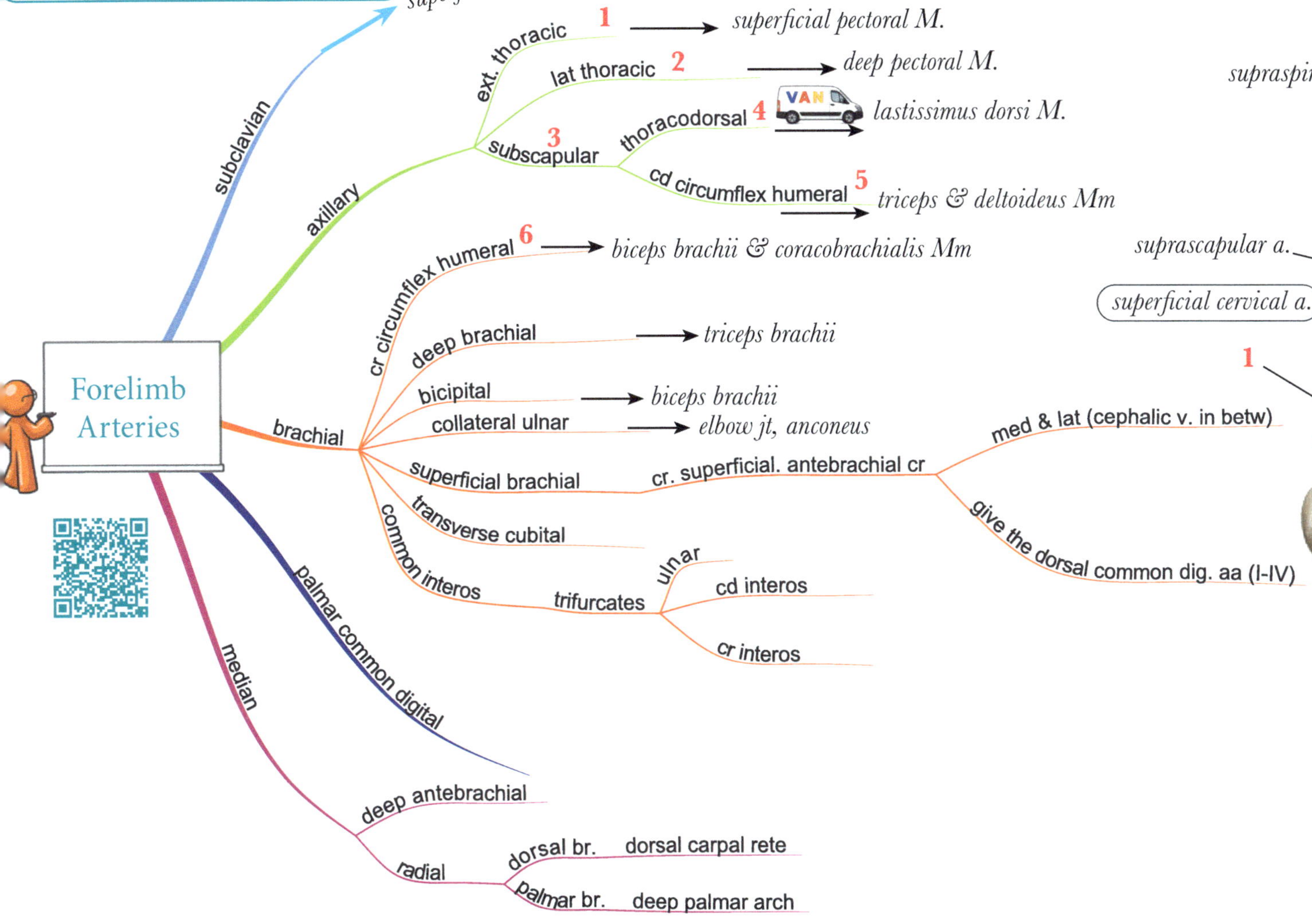

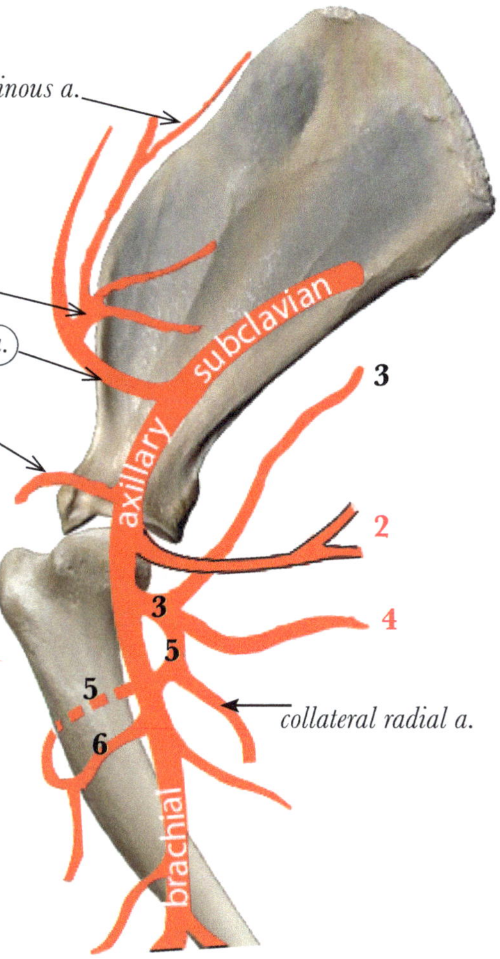

Ulnar + Cd interosseous + Radial → Dorsal carpal rete → Dorsal MC aa. (I-IV)

Arteries of Forelimb

Note:

- *The median n. accompanies the brachial a.*
- *The radial n. accompanies the deep brachial a. in the musculospiral groove*
- *The ulnar n. accompanies the collateral ulnar a.*

Hemorrhage *below the elbow may be controlled by direct pressure on the brachial a.*

in the cat the brachial a. runs through the supracondylar foramen of the humerus together with the median nerve.

in the cat the radial artery is the direct continuation of the median artery

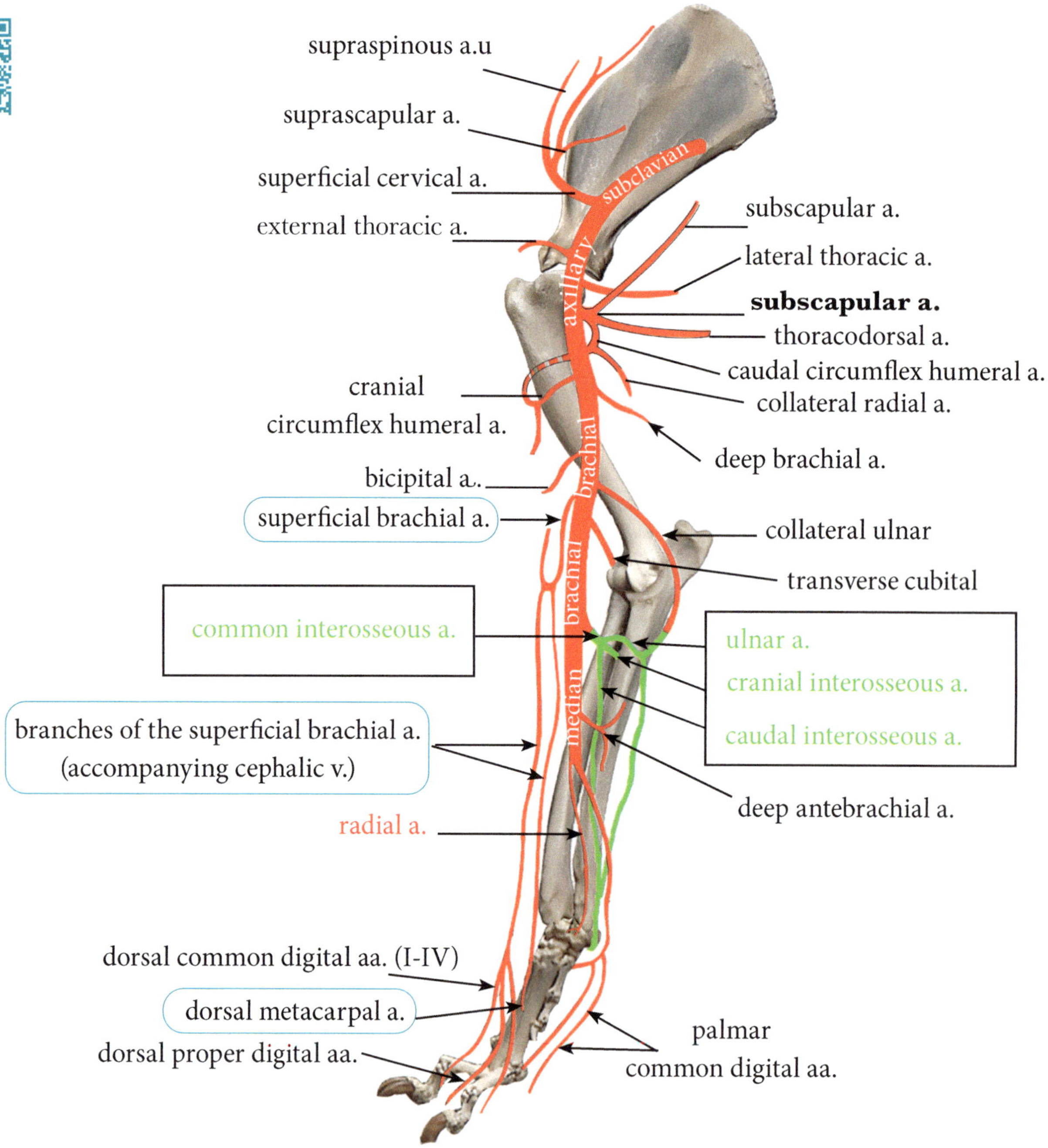

Forelimb Arteries (cont.)

- name according to:
  - surface
    - dorsal (eight Aa)
    - palmar (eight Aa)
  - depth
    - two sets/surface
      - superficial
        - Common dig
          - dorsally
            - dorsal to the ext. tendons
          - palmarly
            - between SDFs & DDFs
      - deep
        - Metacarpals
          - on MC bones
          - deep to flexor tendons (DDF & SDF)
    - four Aa/set
  - position between MCs
    - from med to lat start @ I - IV
    - acc. to the lower # MC
- Aa of MC region
  - dorsally
    - dorsal common dig Aa I-IV
    - dorsal MC Aa I-IV
      - very small
      - from dorsal carpal **rete**
  - palmarly
    - arise from palmar **arches**
    - palmar common dig Aa I - V
      - from superficial palmar arch
    - palmar MC Aa I - IV
      - smaller
      - from deep palmar arch
- Aa of dig region
  - Aa on dorsal & palmar (unlike MC, there is only one vessel on each surface of each digit)
  - @ MCphalangeal jt
    - dorsally
      - common dig Aa join MC Aa
        - proper dig Aa
          - axial
          - abaxial
    - palmarly
      - common dig. Aa join MC Aa
        - proper dig Aa
          - axial
          - abaxial

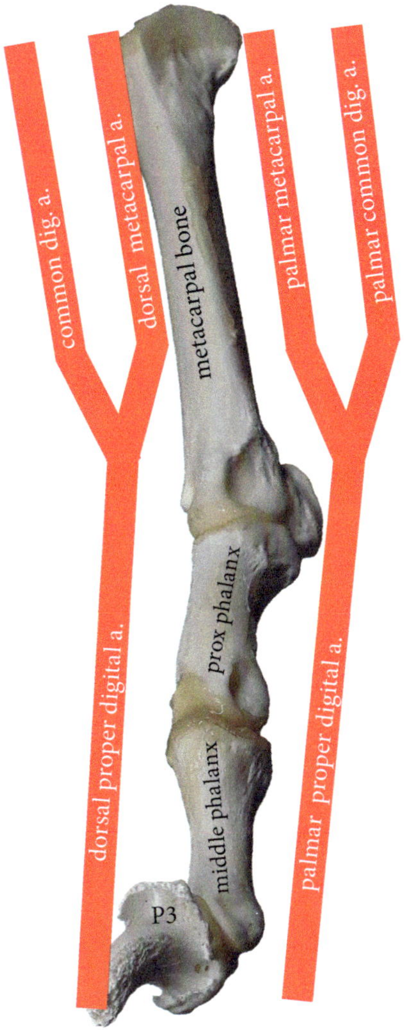

In the MC region, the superficial vessels are named common digital aa & vv, while the deep ones are named MC aa & vv

Veins of Forelimb

- *Superficial set of Vv.*
  - not related to Aa
  - Subcutaneous
  - Dorsal common dig Vv. — form → Accessory cephalic V.
  - Accessory cephalic V.
  - Cephalic V.
    - major superficial V. of paw & forearm
    - veni puncture
    - from sup. palmar venous arch
    - parts
      - antebrachial
      - brachial — drains into ext. jugular v.
  - Median cubital V.
    - joins median & cephalic (antebrachial part)
  - Axillobrachial V.
    - continuation of cephalic V.
    - ends into axillary V.
  - Omobrachial V.
    - br. of axillobrachial V.
    - ends into ext. jugular V.

- **Deep set**
  - parallel Aa
  - brachial v.
  - subclavian v — cd. v. cava

Q: Why can cephalic vein venipuncture be painful in the dog?

A: The cephalic vein is flanked by sensory branches of the superficial radial n. Needle contact with these nerves may cause the dog to cry out and pull the limb away.

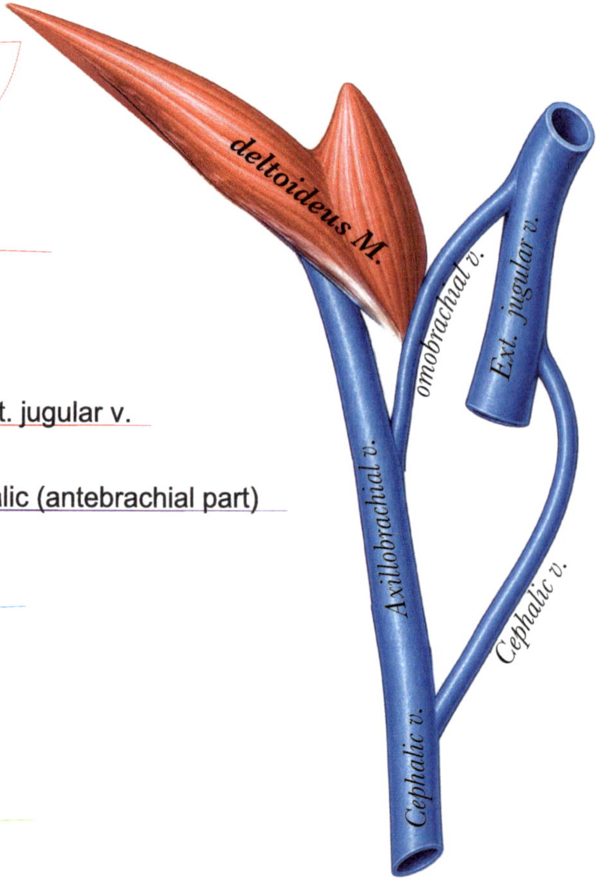

Veins of the forelimb

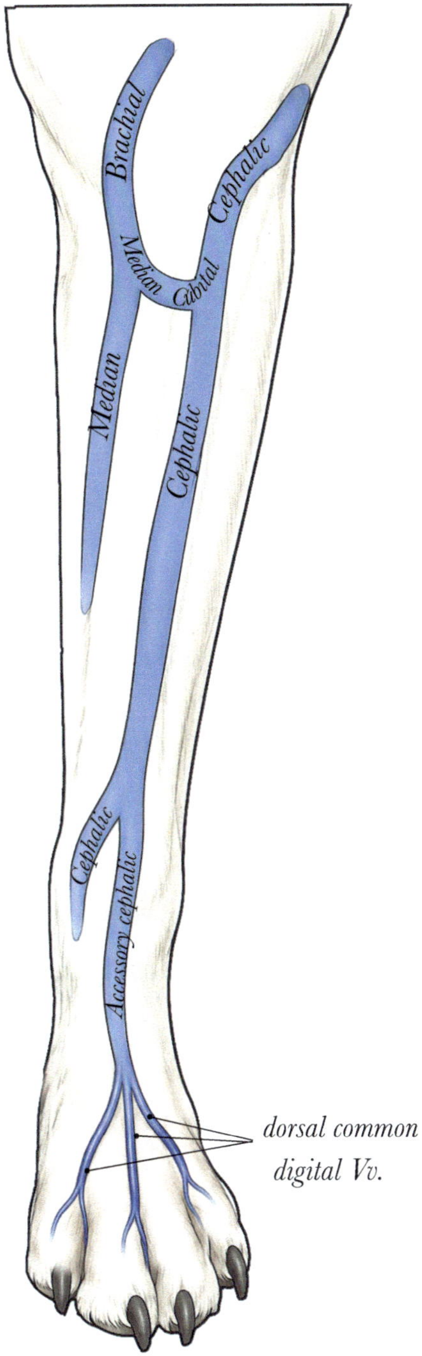

Veins of Forelimb

The Median Cubital v:

*During cephalic vein occlusion (e.g., venipuncture or IV catheter placement), connects the **superficial** venous set (cephalic v.) to the **deep** set (median v.), providing a collateral pathway for blood to be diverted through the median cubital vein into the median vein.*

Lacerations of the paw → laceration of the superficial palmar venous arch → bleeding.

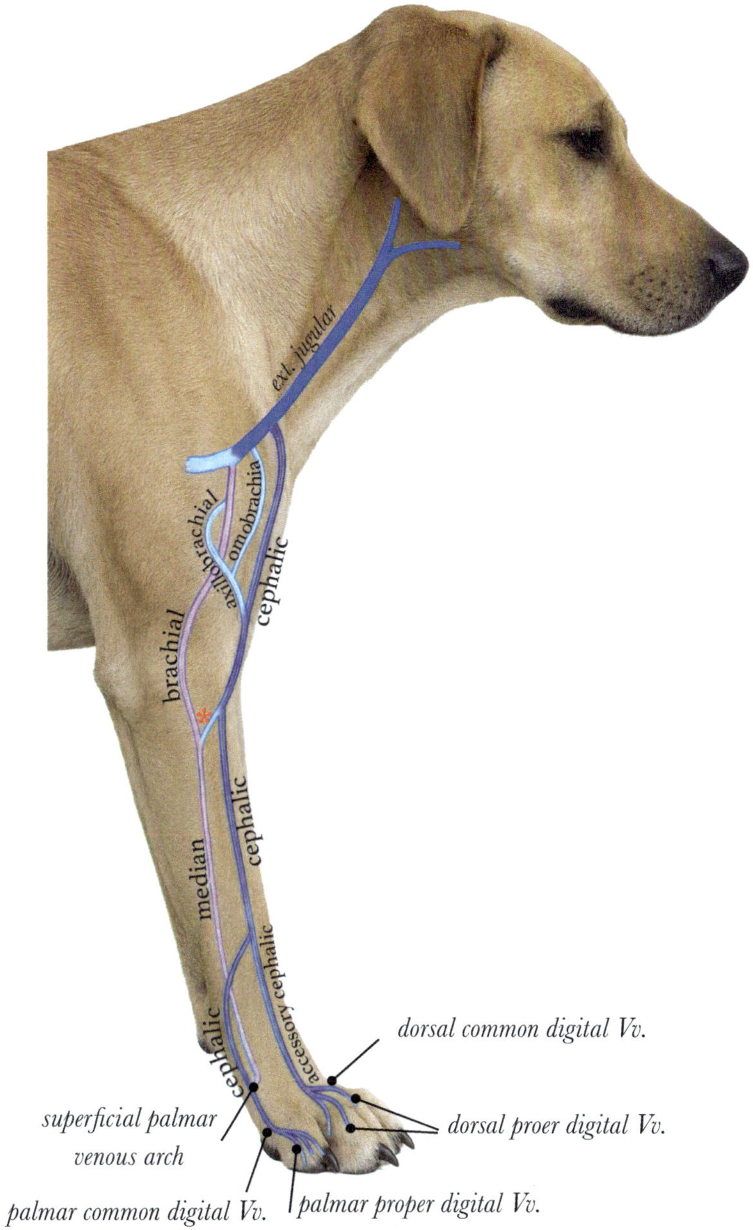

Lymphatics of Forelimb

Two Lymph Centers (LC)

- **Axillary LC**
  - Lnn.
    - axillary Ln.
      - constant
      - caudomedial to shoulder jt.
      - related to rib 2
      - Rt & Lt. communicate
      - metastasis of mammary adenocarcinoma
    - accessory axillary Ln.
      - inconstant
      - cd to axillary Ln
      - @ region of rib 3

      "Accethreey lnn."
      @ region of rib 3 (three)
  - afferent
    - thoracic limb
    - mammary gland
    - thoracic wall
  - efferent
    - thoracic duct
    - lymphatic ducts
    - ext. jugular v.
- **Superficial cervical LC**
  - superficial cervical Lnn.
    - afferent
      - whole forelimb
      - head & neck
    - efferent
      - lymphatic duct (Rt. or Lt.)

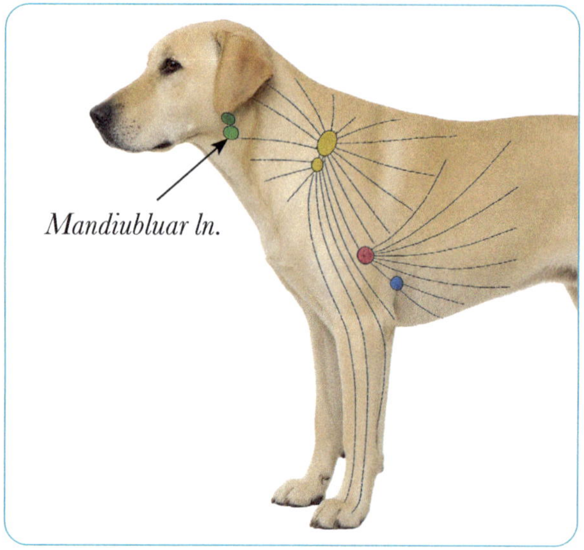

- *Superficial cervical (prescapular) lnn.*
- *Axillary ln. (in medial position)*
- *Accessory axillary ln.*

These are palpable lymph nodes, i.e., they can be detected by touch, especially when enlarged due to infection or cancer.

Nerves of Forelimb

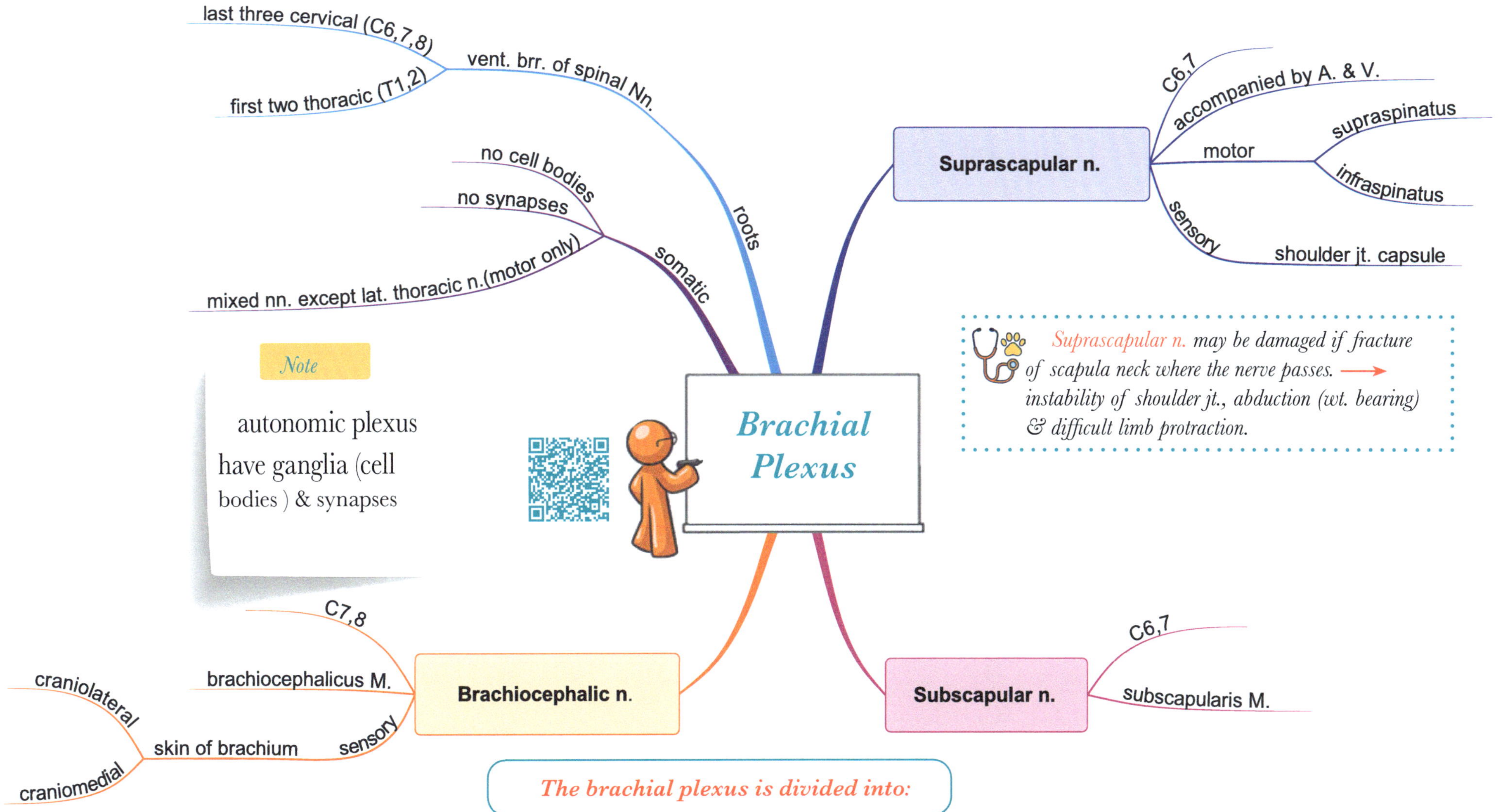

The brachial plexus is divided into:

- *Cranial plexus (C6–C7)* → *the musculocutaneous, axillary, brachiocephalic, suprascapular, subscapular, cr. pectoral & long thoracic nerves.*
- *Caudal plexus (C8–T2)* → *the radial, ulnar, median, lateral thoracic, dorsal thoracic & caudal pectoral nerves.*
- *This division helps localize brachial plexus injuries.e.g. caudal brachial plexus avulsion* → *loss of sensation on caudal antebrachium*

Remember *to use the* ***muscles as landmarks*** *to identify the individual nerves of the brachial plexus.*

Brachial Plexus (cont.)

- **Musculocutaneous n.**
  - C6,7,8
  - BBC Nerve: B B C
  - branches
    - motor
      - Biceps brachii
      - Brachialis
      - Coracobrachialis
      - (Cranial brachial Mm.)
    - sensory
      - med cutaneous antebrachial n.
      - elbow (Cubital) jt. capsule
- **Axillary n.**
  - C7,8
  - betw subscapularis & teres major Mm.
  - branches
    - motor
      - deltoideus, teres major, subscapularis
    - sensory
      - shoulder jt capsule
      - craniolateral cutaneous brachial n.
        - then continues as
      - cr. cut. antebrachial n.

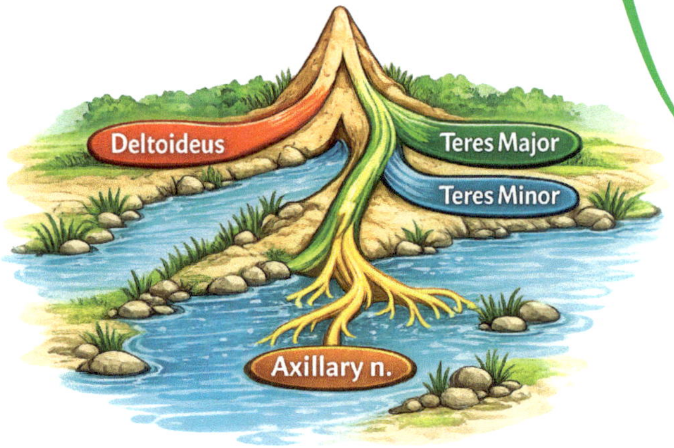

DelTA = ***Del****toideus* + ***T****eres minor* + ***T****eres major* + ***A****nterior (cranial) cutaneous antebrachium* ***(A****xillary n.****)***

Imagine the Axillary n. like a river supplying the ***DelTA***

> ***Autonomous zone (AZ):***
>
> - *An area of skin supplied exclusively by one nerve.*
> - *Loss of sensation in an AZ helps pinpoint the injured peripheral nerve and assists in diagnosing spinal cord lesions.*

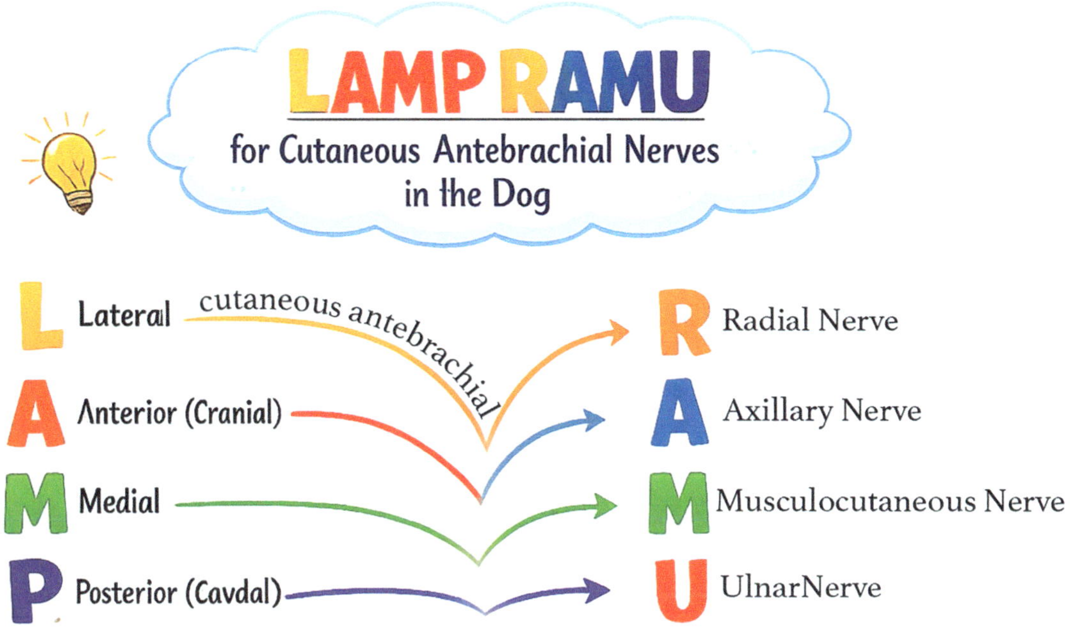

Damage to the Axillary n.

- ***Sensory deficits*** → *lack of sensation on the brachium laterally.*
- ***Motor deficits****: no obvious change to the gait because:* ***Why?***

  - Paralysis of the teres major & minor & deltoid (shoulder flexors) is compensated by long head of triceps brachii.

  - The subscapularis still innervated by the subscapular n.
- *Obvious sign of axillary n. damage is a prominent acromial process due to atrophy of the deltoid M.*

Damage to the Musculocutaneous n. :

- *Atrophy: Biceps & brachialis Mm.*
- *Elbow flexion: Weak, but partly compensated by extensor carpi radialis, supinator & pronator Mm.*
- *Sensory loss: Medial antebrachium.*

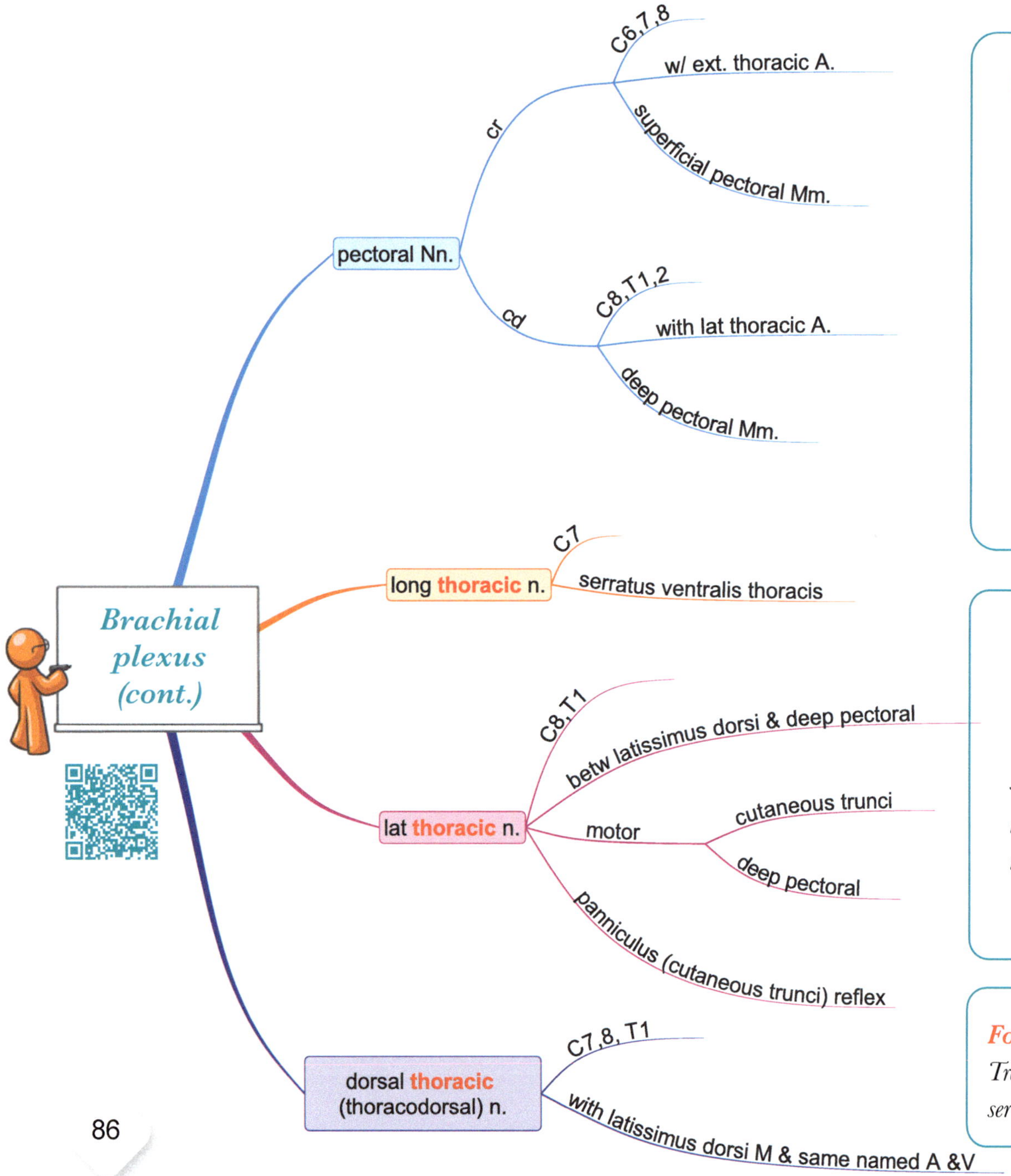

Complete avulsion (abduction) of the brachial plexus *as in car accedents* ⟶

- *Paralysis of Mm, dropped shoulder, elbow & carpal flexion.*
- *Loss of the panniculus reflex.*
- *Injury of the sympathetic preganglionics to cervicothoracic ganglion* ⟶ *ipsilateral Horner's syndrome.*
- *Sensory deficits in forelimb (but skin over long head of triceps brachii will be normal as it is supplied by the intercostobrachial nn).*

The panniculus (cutaneous trunci) reflex:

Used to localize spinal cord lesions.

- *Pinching the skin between T1 & L4* ⟶ *sensory signals travel via dorsal root ganglia (DRG)* ⟶ *propriospinal tract* ⟶ *synapse at motor neurons (C8–T1) that form the lat. thoracic n.* ⟶ *cutaneous trunci M.* ⟶ *twitching of skin.*

Forelimb structures not supplied by brachial plexus: *Trapezius, omotransversarius, part of brachiocephalicus, rhomboideus, serratus ventralis cervicis, & skin of prox. shoulder.*

Radial n.

- C7,8, T1,2
- betw med & long heads of triceps
  - in brachialis groove (site of injury)

- deep to lat head of triceps divides into:
  - both superficial & deep brs continue to the paw

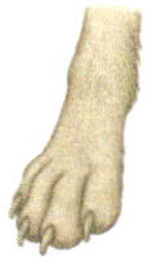

- **Deep br.** (motor)
  - deep to ext. carpi radialis
    - tensor fascia antebrachii
    - triceps brachii (all heads)
    - anconeus
  - extensors of carpus & digit
    - **E**xt. carpi radialis
    - **C**ommon dig. ext
    - **L**at dig. extensor
    - **U**lnaris lateralis
    - abductor pollicis longus
    - supinator
- **Superficial br.** (sensory)
  - along suprficial brachial A. & cephalic V.
  - skin of cr. forearm & dorsum paw
  - avoid in cephalic venipuncture
  - capsule of carpus & MC jts.
  - overlap with brs of axiallary n.
  - lat br.
    - lateral cut. antebrachial n.
    - dorsal common digital n. II, III & IV
  - medial br.
    - dorsal digital n. I
    - abaxial dorsal proper digital n. II

What happens if a nerve is damaged?

Ask yourself the following questions:

Which muscles does the nerve supply?

What is the action of these muscles?

Which movements will be affected?

Can the animal still stand and bear weight?

Does this nerve have an autonomous zone (AZ) where sensory loss can be reliably detected?

Avoid the ***Radial n.*** *during lat. approach to the humerus as it runs in the musculospiral groove.*

Distal radial n. paralysis (fracture of distal humerus where the n. passes in the brachialis groove or over the lat. supracondylar crest ⟶ paralysis of carpal & digital extensors --> "Kiss hand Position" or dropped wrist in human.

Prox. radial n. paralysis ⟶ paralysis of extensors of ***elbow****, carpus & dig ⟶ animal can't bear weight + prominent cd border of scapula + olecranon (****atrophy of triceps brachii****).*

Sensory deficits in the dorsal aspect of the paw.

Radial n. paralysis in Rt. forelimb

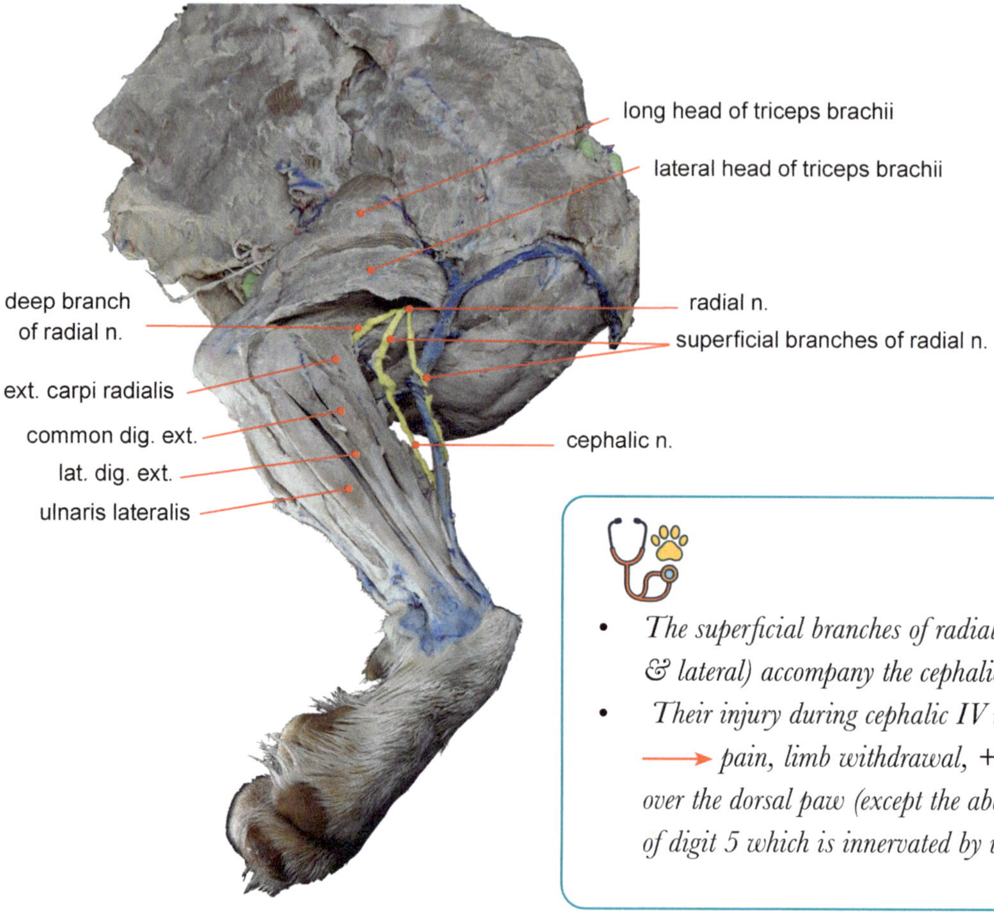

- *The superficial branches of radial n. (medial & lateral) accompany the cephalic vein.*
- *Their injury during cephalic IV injections ⟶ pain, limb withdrawal, + sensory loss over the dorsal paw (except the abaxial surface of digit 5 which is innervated by ulnar n.*

The radial, axillary, and suprascapular nerves pass from medial side where the brachial plexus to the lateral side of the limb.

- *Most muscles of the paw supplied by ulnar n.*
- ***Digit V sensory innervation:***
  - *palmar aspect by ulnar n. (both sides axial & abaxial).*
  - *dorsal aspect by ulnar n.(abaxial surface) & radial → axial*

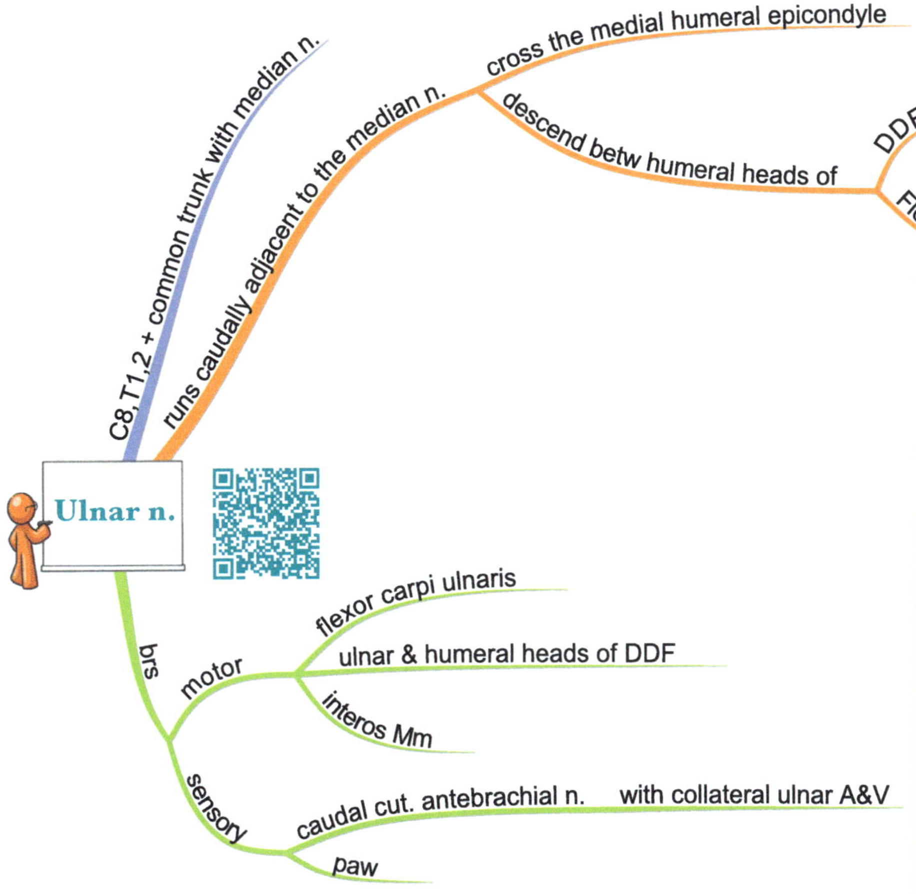

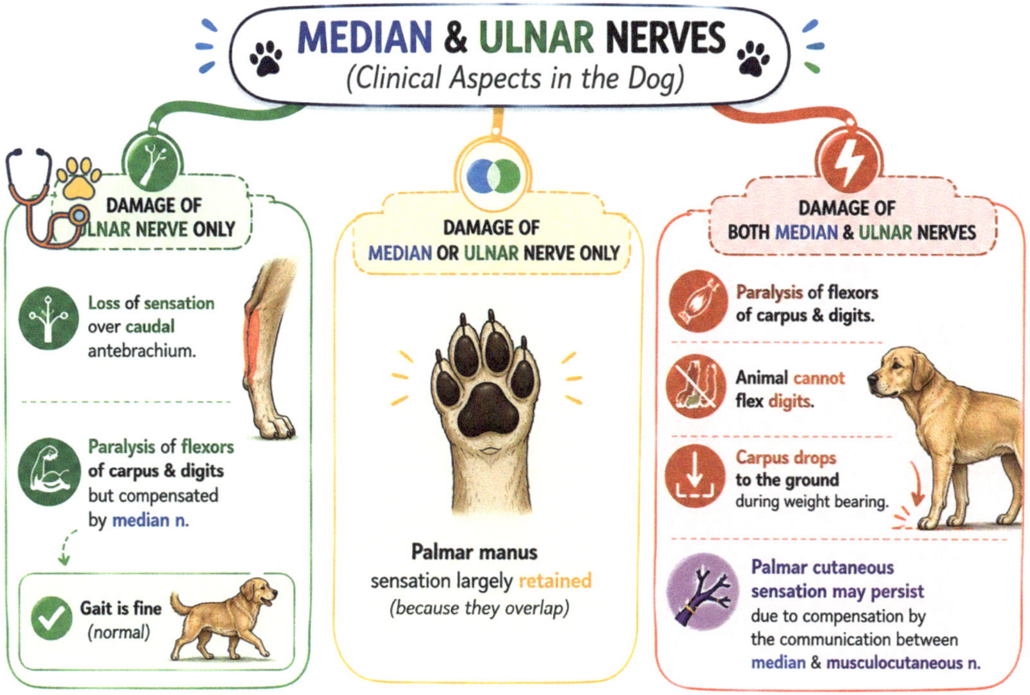

Median nerve

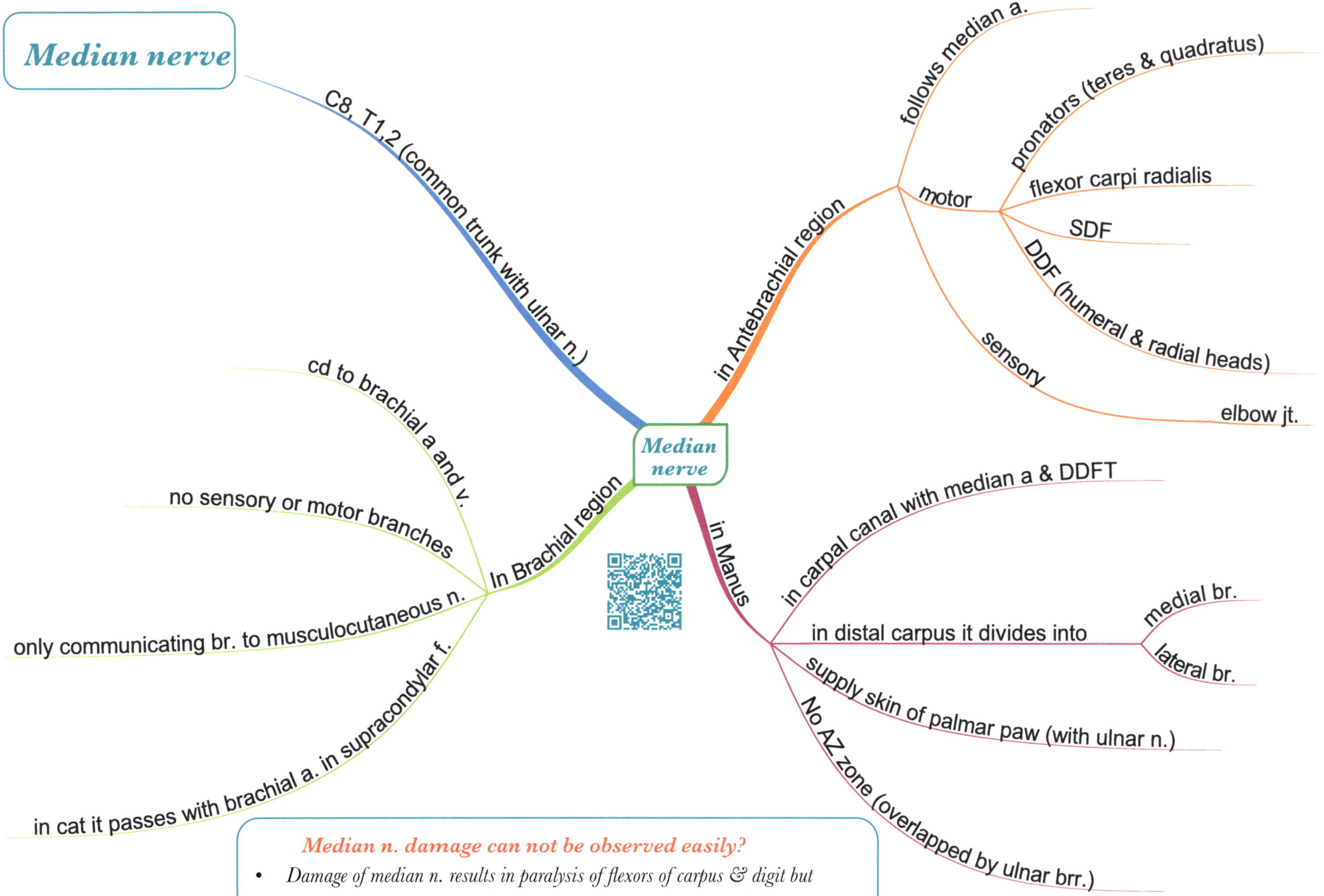

Median n. damage can not be observed easily?

- *Damage of median n. results in paralysis of flexors of carpus & digit but compensated by ulnar n which supplies some of the flexors,so gait looks fine.*
- *No cutaneous sensory deficits, as the median n. has no autonomous zone (AZ).*

Nerves in the Manus

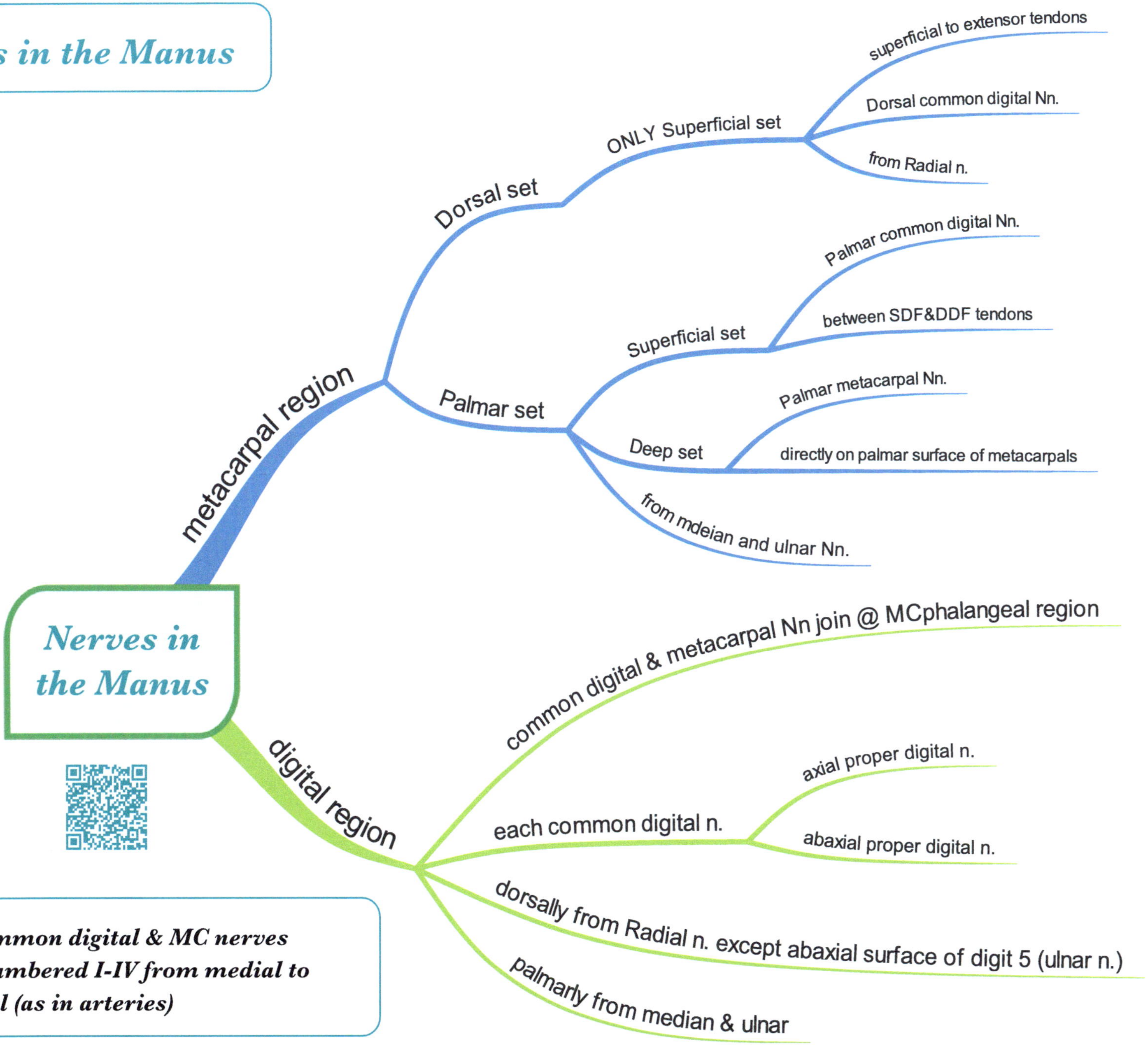

The common digital & MC nerves are numbered I-IV from medial to lateral (as in arteries)

Nerves in the manus

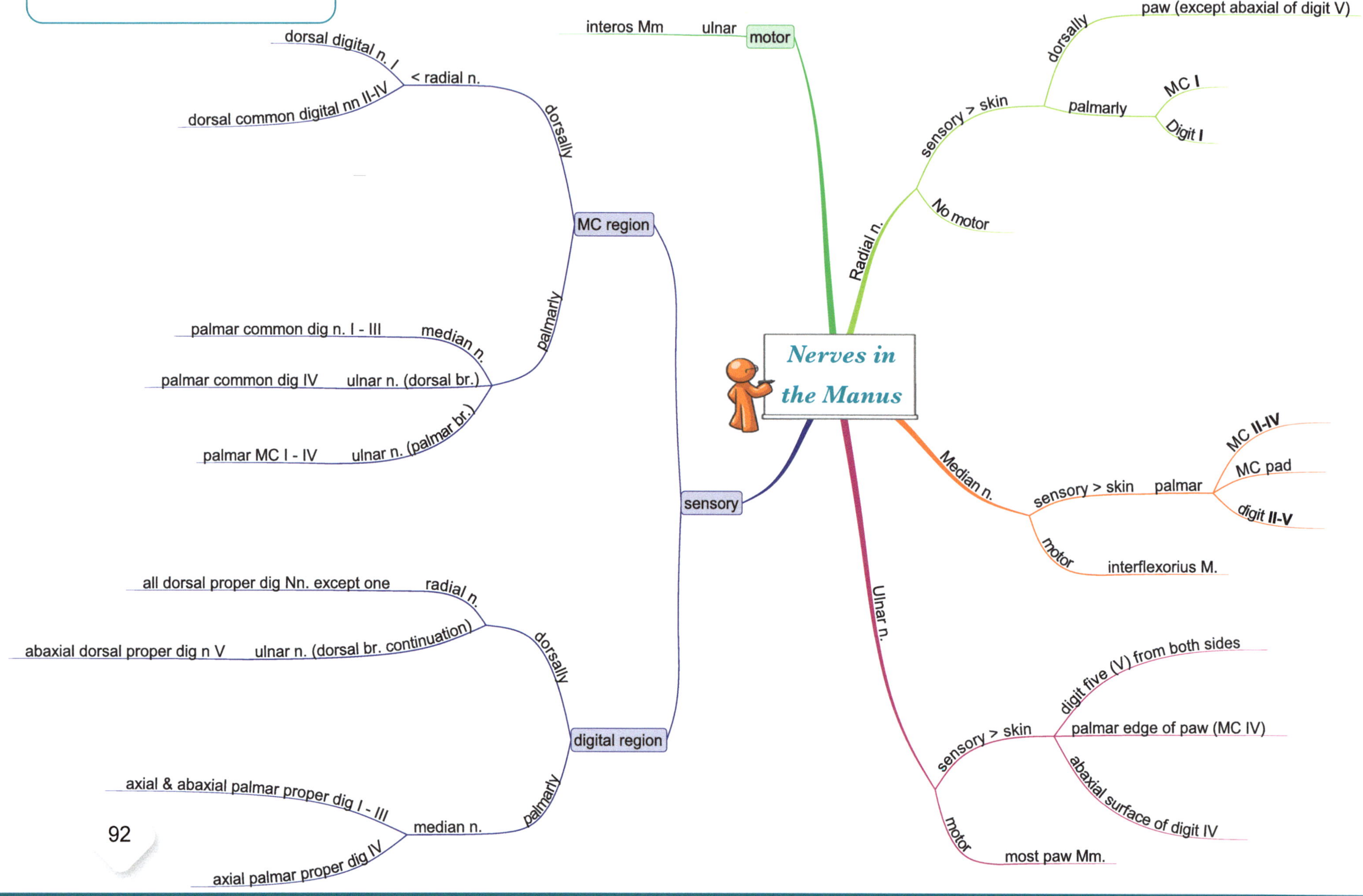

QR codes Appendix

Page	Audio QR	Topic	Transcript
10		Planes	
10		Directional	
10		Movement	
13		Skeleton	
16		Types of bones (shape)	
18		Bony projections	
20		Bony depressions and openings	
23		Joint types	
24		Synovial joint structure	
25		Synovial joint types	
26		Naming muscles	
28		Fascia	
29		Compartmental syndrome	
30		Tendon and aponeurosis	
33		Forelimb bones	
34		Scapula	

QR codes Appendix

Page	Audio QR	Topic	Transcript
36		Humerus	
38		Radius	
40		Ulna	
43		Forelimb bones clinical insights	
44		Manus	
46		Digits I-V	
49		Shoulder joint	
50		Elbow joint	
52		Carpal joint	
54		Metacarpophalangeal joint	
55		Interphalangeal joints	
58		Synsarcosis	
60		Shoulder muscles	
63		Brachium muscles	
66		Forearm muscles Craniolateral group	
70		Forearm muscles Caudomedial group	

QR codes Appendix

Page	Audio QR	Topic	Transcript
72		Deep Digital Flexor muscle	
73		Carpal canal	
74		Forelimb muscle actions	
77		Arteries of shoulder	
78		Arteries of bracium & antebrachium	
79		Arteries of forelimb 3	
80		Veins of forelimb	
82		Lymphatics of forelimb	
83		Brachial plexus Suprascapular n Supbscapular n Brachiocephalic n	
84		Musculocut. n. Axillary n.	
86		Pectoral Nn. Long thoracic n. Lat. thoracic n Dorsal thoracic n.	
87		Radial nerve	
89		Ulnar nerve	
90		Median nerve	
91		Nerves in the manus	